Introduction to
Classical Mechanics
Solutions to Problems

Introduction to
Classical Mechanics
Solutions to Problems

John Dirk Walecka
College of William and Mary, USA

World Scientific

NEW JERSEY · LONDON · SINGAPORE · BEIJING · SHANGHAI · HONG KONG · TAIPEI · CHENNAI · TOKYO

Published by

World Scientific Publishing Co. Pte. Ltd.
5 Toh Tuck Link, Singapore 596224
USA office: 27 Warren Street, Suite 401-402, Hackensack, NJ 07601
UK office: 57 Shelton Street, Covent Garden, London WC2H 9HE

Library of Congress Cataloging-in-Publication Data
Names: Walecka, John Dirk, 1932–   author.
Title: Introduction to classical mechanics : solutions to problems / John Dirk Walecka,
    College of William and Mary, USA.
Description: New Jersey : World Scientific, [2021] | Includes bibliographical
    references and index.
Identifiers: LCCN 2020034547 | ISBN 9789811224942 (hardcover) |
    ISBN 9789811227622 (paperback) | ISBN 9789811224959 (ebook)
Subjects: LCSH: Mechanics--Textbooks.
Classification: LCC QC125.2 .W263 2021 | DDC 531--dc23
LC record available at https://lccn.loc.gov/2020034547

British Library Cataloguing-in-Publication Data
A catalogue record for this book is available from the British Library.

For any available supplementary material, please visit
https://www.worldscientific.com/worldscibooks/10.1142/11955#t=suppl

Printed in Singapore

For Kay

Preface

The author recently published a book entitled *Introduction to Electricity and Magnetism* [Walecka (2018)]. It is based on an introductory course taught several years ago at Stanford, with over 400 students enrolled. The only requirements were an elementary knowledge of calculus and familiarity with vectors and Newton's laws; the development was otherwise self-contained. The lectures, although relatively concise, take one from Coulomb's law to Maxwell's equations and special relativity in a lucid and logical fashion. The book has an extensive set of accessible problems that enhances and extends the coverage. As an aid to teaching and learning, the solutions to those problems were subsequently published in a separate text [Walecka (2019)].

Although never presented in an actual course, it occurred to the author that it would be fun to compose an equivalent set of lectures, aimed at the very best students, that would serve as a *prequel* to that *Electricity and Magnetism* text. These lectures would assume a good, concurrent, course in calculus and familiarity with basic concepts in physics (say, from a good high-school course); they would otherwise, again, be self-contained. For my own amusement, I did just that.

The lectures start with a review of the necessary mathematics and a review of vectors. The idea of an inertial frame is then introduced, and Newton's laws are stated, with several applications included. The concepts of energy and angular momentum are introduced, and the analysis is then extended to many-particle systems.

The notions of generalized coordinates and Lagrange's equations are first introduced on the basis that they reproduce Newton's laws in the chosen examples. After a lecture introducing the *calculus of variations*, Lagrange's equations are derived from what then serves as the basic principle of classical mechanics —*Hamilton's principle of stationary action.* Several more examples are given of lagrangian mechanics.

Hamilton's equations are similarly first introduced on the basis that they reproduce Lagrange's equations and Newton's laws for the chosen examples, and they are then subsequently similarly derived from Hamilton's principle of stationary action. Several examples are included of hamiltonian mechanics and phase space.

A lecture then discusses the transition from the mechanics of discrete particle systems to that of *continuous media.* Lagrange's equations for continuous systems are exhibited and then derived from Hamilton's principle. The wave motion of a string under tension serves as the paradigm for continuum mechanics, and the analysis extends up through the construction of the energy-momentum tensor and the reflection and radiation of those waves.

Irrotational, isentropic fluid flow, where the velocity field is derived from a potential and there is no internal (reversible) heat flow, serves as the final example of lagrangian continuum mechanics. The lagrangian density is constructed. Bernoulli's equation and the continuity equation for the mass (number) density are then derived from Lagrange's equations, and they are related back to Newton's laws for fluid mechanics. The energy density and energy flux are constructed, and the analysis is then applied to sound waves, where reflection and radiation are again examined.

These lectures have now been published as the book *Introduction to Classical Mechanics* [Walecka (2020)]. The *goal* of this text is to provide a clear and concise set of lectures that take one from the introduction and application of Newton's laws up to Hamilton's principle and the lagrangian mechanics of continuous systems. This, indeed, provides the point of departure from classical mechanics to modern quantum field theory.[1] An extensive set of accessible problems again enhances and extends the coverage. As was previously the case for the electromagnetism texts, this additional book provides the solutions to those classical mechanics problems.

I would like to once again thank my editor, Ms. Lakshmi Narayanan, for her help and support on this project.

Williamsburg, Virginia   *John Dirk Walecka*
June 17, 2020   *Governor's Distinguished CEBAF*
                    Professor of Physics, emeritus
                    College of William and Mary

---

[1] See, for example, [Walecka (2010)].

Contents

Chapter 1

Introduction

Problem 1.1 Use the definition in Eq. (1.2) to derive the following derivatives

$$\frac{d}{dt}(t^n) = nt^{n-1} \qquad ; \text{ integer } n$$

$$\frac{d}{d\theta}(\sin\theta) = \cos\theta$$

$$\frac{d}{d\theta}(\cos\theta) = -\sin\theta$$

Solution to Problem 1.1

For the first relation, write

$$(t + \Delta t)^n = t^n + n(\Delta t)t^{n-1} + O[(\Delta t)^2]$$

Then

$$\frac{1}{\Delta t}\left[(t + \Delta t)^n - t^n\right] = nt^{n-1} + O(\Delta t)$$

Now take the limit $\Delta t \to 0$.

For the second relation, write[1]

$$\sin(\theta + \Delta\theta) = \sin\theta\cos(\Delta\theta) + \cos\theta\sin(\Delta\theta)$$
$$= \sin\theta + (\Delta\theta)\cos\theta + O[(\Delta\theta)^2]$$

Then

$$\frac{1}{\Delta\theta}\left[\sin(\theta + \Delta\theta) - \sin\theta\right] = \cos\theta + O(\Delta\theta)$$

---

[1]It is assumed that the reader knows these trigonometric identities and expansions; they follow from Probs. (1.7)–(1.8).

Now take the limit $\Delta\theta \to 0$.

Similarly, for the third relation,

$$\cos(\theta + \Delta\theta) = \cos\theta\cos(\Delta\theta) - \sin\theta\sin(\Delta\theta)$$
$$= \cos\theta - (\Delta\theta)\sin\theta + O[(\Delta\theta)^2]$$

Then

$$\frac{1}{\Delta\theta}[\cos(\theta + \Delta\theta) - \cos\theta] = -\sin\theta + O(\Delta\theta)$$

Again, take the limit $\Delta\theta \to 0$.

Problem 1.2 Use the relations in Eqs. (1.5)–(1.7) to verify the following integrals

$$\int_0^t t^n\, dt = \frac{1}{n+1}t^{n+1}$$

$$\int_0^\theta (\cos u)\, du = \sin\theta$$

$$\int_0^\theta (\sin u)\, du = 1 - \cos\theta$$

Solution to Problem 1.2

For the first relation, use the first result from Prob. 1.1 to write[2]

$$\int_0^t t^n\, dt = \left[\frac{1}{n+1}t^{n+1}\right]_0^t = \frac{1}{n+1}t^{n+1}$$

For the second relation, use the second result from Prob. 1.1 to write

$$\int_0^\theta (\cos u)\, du = [\sin\theta]_0^\theta = \sin\theta$$

Similarly, for the third relation,

$$\int_0^\theta (\sin u)\, du = [-\cos\theta]_0^\theta = 1 - \cos\theta$$

Problem 1.3 Show that for the derivative of a product

$$\frac{d}{dx}[f(x)g(x)] = \frac{df(x)}{dx}g(x) + f(x)\frac{dg(x)}{dx}$$

---

[2]The notation is $[f(t)]_a^b \equiv f(t_b) - f(t_a)$.

Solution to Problem 1.3

Write

$$[f(x + \Delta x)] [g(x + \Delta x)]$$
$$= \left[f(x) + (\Delta x)\frac{df(x)}{dx} + O[(\Delta x)^2] \right] \left[g(x) + (\Delta x)\frac{dg(x)}{dx} + O[(\Delta x)^2] \right]$$
$$= f(x)g(x) + (\Delta x)\left[\frac{df(x)}{dx}g(x) + f(x)\frac{dg(x)}{dx} \right] + O[(\Delta x)^2]$$

The result now follows.[3]

Problem 1.4 (a) Differentiate the relation $t^{1/2}t^{1/2} = t$ to show that

$$\frac{d}{dt}t^{1/2} = \frac{1}{2}t^{-1/2}$$

(b) Generalize this result to conclude that the first of Eqs. (17.1) holds for all integer and half-integer n.

Solution to Problem 1.4

(a) Take the derivative of the given relation, and use the result from the previous problem

$$\frac{d}{dt}\left(t^{1/2}t^{1/2} \right) = 2t^{1/2}\frac{d}{dt}\left(t^{1/2} \right) = 1$$

Hence

$$\frac{d}{dt}\left(t^{1/2} \right) = \frac{1}{2\,t^{1/2}}$$

Similarly

$$\frac{d}{dt}\left(t^{1/2}t^{-1/2} \right) = t^{1/2}\frac{d}{dt}\left(t^{-1/2} \right) + t^{-1/2}\frac{d}{dt}\left(t^{1/2} \right)$$
$$= t^{1/2}\frac{d}{dt}\left(t^{-1/2} \right) + t^{-1/2}\left(\frac{1}{2t^{1/2}} \right) = 0$$

Hence

$$\frac{d}{dt}\left(t^{-1/2} \right) = -\frac{1}{2\,t^{3/2}}$$

Thus the first of Eqs. (17.1) also holds for $n = \pm 1/2$.

---

[3]Note that we deal with two types of differences in this work. There is the finite difference Δx, and then the *differential dx* that becomes vanishingly small [recall Eqs. (1.2) and (2.26)].

(b) To start extending these results, differentiate, for example,

$$\frac{d}{dt}\left(t^{3/2}t^{1/2}\right) = t^{1/2}\frac{d}{dt}\left(t^{3/2}\right) + t^{3/2}\frac{d}{dt}\left(t^{1/2}\right)$$

$$= t^{1/2}\frac{d}{dt}\left(t^{3/2}\right) + t^{3/2}\left(\frac{1}{2t^{1/2}}\right) = 2t$$

Hence

$$\frac{d}{dt}\left(t^{3/2}\right) = \frac{3}{2}t^{1/2}$$

and so on. It follows, in this manner, that the first of Eqs. (17.1) holds for all integer and half-integer n.

Problem 1.5[4] Integrate the relation in Prob. 1.3 between two points (a, b)

$$\int_a^b dx\,\frac{d}{dx}\left[f(x)g(x)\right] = \int_a^b dx\,\frac{df(x)}{dx}g(x) + \int_a^b dx\,f(x)\frac{dg(x)}{dx}$$

Since the integrand of the first term is a perfect differential, this relation is rewritten as

$$[f(x)g(x)]_a^b = \int_a^b dx\,\frac{df(x)}{dx}g(x) + \int_a^b dx\,f(x)\frac{dg(x)}{dx}$$

If the boundary conditions are such that this first term *vanishes*, then

$$\int_a^b dx\,\frac{df(x)}{dx}g(x) = -\int_a^b dx\,f(x)\frac{dg(x)}{dx} \qquad ;\; \text{if } [f(x)g(x)]_a^b = 0$$

This is known as *partial integration*.

Solution to Problem 1.5

From Prob. 1.3, the derivative of a product is

$$\frac{d}{dx}\left[f(x)g(x)\right] = \frac{df(x)}{dx}g(x) + f(x)\frac{dg(x)}{dx}$$

Integrate this expression between two points (a, b). The integral of the derivative of a function is the difference of the function evaluated at the two endpoints. Thus

$$\int_a^b dx\,\frac{d}{dx}\left[f(x)g(x)\right] = [f(x)g(x)]_{x=x_b} - [f(x)g(x)]_{x=x_a}$$

$$\equiv [f(x)g(x)]_a^b$$

---

[4]The relations in Probs. 1.5–1.6 play a central role in our analysis.

Now suppose the *boundary conditions* in the problem are such that this expression vanishes at the endpoints, so that $[f(x)g(x)]_a^b = 0$. It follows that then

$$\int_a^b dx \, \frac{df(x)}{dx} g(x) = -\int_a^b dx \, f(x) \frac{dg(x)}{dx} \qquad ; \text{ if } [f(x)g(x)]_a^b = 0$$

This is known as *partial integration*. It is an extremely valuable tool.

Problem 1.6 Derive the *chain rule*

$$\frac{d}{dx} f[u(x)] = \frac{df(u)}{du} \frac{du(x)}{dx}$$

Solution to Problem 1.6

Write

$$f[u(x + \Delta x)] = f \left\{ u(x) + (\Delta x) \frac{du(x)}{dx} + O\left[(\Delta x)^2\right] \right\}$$

$$= f[u(x)] + \left\{ (\Delta x) \frac{du(x)}{dx} + O\left[(\Delta x)^2\right] \right\} \frac{df(u)}{du} + O\left[(\Delta x)^2\right]$$

$$= f[u(x)] + (\Delta x) \left[\frac{df(u)}{du} \frac{du(x)}{dx} \right] + O\left[(\Delta x)^2\right]$$

The result now follows.

Problem 1.7 The exponential function can be defined by the power series

$$e^x \equiv \sum_{n=0}^{\infty} \frac{x^n}{n!}$$

where $n! = n(n-1)(n-2) \cdots 1$, and $x^0 = 0! = 1$.
  (a) Show

$$\frac{de^x}{dx} = e^x$$

  (b) Let i be the imaginary number $i = \sqrt{-1}$. Show

$$e^{i\theta} = \cos\theta + i\sin\theta$$

where

$$\cos\theta = 1 - \frac{\theta^2}{2!} + \frac{\theta^4}{4!} - \frac{\theta^6}{6!} + \cdots$$

$$\sin\theta = \theta - \frac{\theta^3}{3!} + \frac{\theta^5}{5!} - \frac{\theta^7}{7!} + \cdots$$

Solution to Problem 1.7

(a) Assume there is enough convergence so that the series can be differentiated term-by-term[5]

$$\frac{de^x}{dx} = \sum_{n=0}^{\infty} \frac{n x^{n-1}}{n!} = \sum_{n=1}^{\infty} \frac{x^{n-1}}{(n-1)!}$$

Now introduce $m \equiv n - 1$

$$\frac{de^x}{dx} = \sum_{m=0}^{\infty} \frac{x^m}{m!} = e^x$$

(b) Use

$$i^2 = -1 \qquad ; \; i^4 = 1$$

Then

$$e^{i\theta} = 1 + i\theta - \frac{\theta^2}{2!} - i\frac{\theta^3}{3!} + \frac{\theta^4}{4!} + i\frac{\theta^5}{5!} - \frac{\theta^6}{6!} - i\frac{\theta^7}{7!} + \cdots$$

This is rewritten as

$$e^{i\theta} = \cos\theta + i\sin\theta$$

where

$$\cos\theta \equiv 1 - \frac{\theta^2}{2!} + \frac{\theta^4}{4!} - \frac{\theta^6}{6!} \cdots$$

$$\sin\theta \equiv \theta - \frac{\theta^3}{3!} + \frac{\theta^5}{5!} - \frac{\theta^7}{7!} + \cdots$$

Problem 1.8 The binomial theorem states that for integer p,[6]

$$(x+y)^p = \sum_{m=0}^{p} \sum_{n=0}^{p} \frac{p!}{m!n!} x^m y^n \qquad ; \; m+n = p$$

Use the binomial theorem to show that

$$e^x e^y = e^{x+y}$$

---

[5] Actually, there is.

[6] The constraint can be explicitly built into the sums by including a factor $\delta_{p,m+n}$ in the summand, where this is the Kronecker delta of Probs. 2.1 and 13.3.

Solution to Problem 1.8

Write

$$e^{x+y} = \sum_{p=0}^{\infty} \frac{(x+y)^p}{p!}$$

Use the binomial theorem in the statement of the problem, along with the statement of the constraint $m + n = p$ in the footnote

$$\sum_{p=0}^{\infty} \frac{(x+y)^p}{p!} = \sum_{p=0}^{\infty} \sum_{m=0}^{p} \sum_{n=0}^{p} \delta_{p,m+n} \frac{1}{m!n!} x^m y^n$$

Now m and n are positive integers (including zero). Summing over all positive integer pairs (m, n) for a given value of their sum $m + n$, and then summing over all values of that sum, is equivalent to just summing over all values of those integers m and n individually.[7] Hence

$$\sum_{p=0}^{\infty} \sum_{m=0}^{p} \sum_{n=0}^{p} \delta_{p,m+n} \frac{1}{m!n!} x^m y^n = \sum_{m=0}^{\infty} \sum_{n=0}^{\infty} \frac{x^m}{m!} \frac{y^n}{n!}$$

It follows that

$$e^{x+y} = e^x e^y$$

****

(*Aside*) Various trigonometric identities follow from this relation. For example, consider

$$e^{i(\theta_1+\theta_2)} = e^{i\theta_1} e^{i\theta_2}$$

This yields the following trigonometric identities

$$\cos(\theta_1 + \theta_2) + i \sin(\theta_1 + \theta_2) = (\cos\theta_1 + i\sin\theta_1)(\cos\theta_2 + i\sin\theta_2) =$$
$$(\cos\theta_1 \cos\theta_2 - \sin\theta_1 \sin\theta_2) + i(\cos\theta_1 \sin\theta_2 + \sin\theta_1 \cos\theta_2)$$

Furthermore, the function $z = e^{i\theta}$ describes the unit circle in the complex plane, and the angle θ, in *radians*, is the arc length on that circle.
****

Problem 1.9 The function $\ln(x)$ is defined by

$$x \equiv e^{\ln(x)}$$

---

[7] Just write out a few terms. More formally, use the presence of the factor $\delta_{p,m+n}$ to extend the sums over (m, n) to infinity, and then use $\sum_{p=0}^{\infty} \delta_{p,m+n} = 1$.

Differentiate this expression with respect to x, to obtain

$$1 = e^{\ln(x)}\left[\frac{d\ln(x)}{dx}\right] = x\left[\frac{d\ln(x)}{dx}\right]$$

Hence, conclude that

$$\frac{d\ln(x)}{dx} = \frac{1}{x}$$

Solution to Problem 1.9

Differentiate the defining relation, and make use of the chain rule in Prob. 1.6 with $u(x) \equiv \ln(x)$,

$$1 = \frac{d}{dx}e^{\ln(x)} = \left[\frac{d}{du}e^u\right]\left[\frac{d}{dx}u(x)\right] \qquad ; \; u(x) = \ln(x)$$

$$= e^u\left[\frac{d}{dx}u(x)\right]$$

$$= x\frac{d}{dx}\ln(x)$$

Hence

$$\frac{d\ln(x)}{dx} = \frac{1}{x}$$

Problem 1.10 (a) Sketch the function

$$y(x) = ax + bx^2$$

where (a, b) are positive constants. Show the *minimum* of the function occurs at the place where the derivative vanishes

$$\frac{dy(x)}{dx} = 0 \qquad ; \text{ for minimum}$$

(b) Now sketch the function

$$y(x) = ax - bx^2$$

Show the *maximum* of the function occurs where the derivative vanishes

$$\frac{dy(x)}{dx} = 0 \qquad ; \text{ for maximum}$$

(b) Sketch the function

$$y(x) = ax^3$$

Show the *point of inflection* of the function occurs where the derivative vanishes

$$\frac{dy(x)}{dx} = 0 \qquad \text{; point of inflection}$$

Solution to Problem 1.10

(a) The curve can be rewritten as

$$y = b\left(x + \frac{a}{2b}\right)^2 - \frac{a^2}{4b}$$

This is a *parabola* that opens upwards, with its vertex at $(x, y) = (-a/2b, -a^2/4b)$. The minimum of this curve occurs at the point where the slope vanishes

$$\frac{d}{dx}\left(ax + bx^2\right) = a + 2bx = 0 \qquad \text{; minimum}$$

(b) Now

$$y = -b\left(x - \frac{a}{2b}\right)^2 + \frac{a^2}{4b}$$

This is a parabola that opens *downwards*, with its vertex at $(x, y) = (a/2b, a^2/4b)$. The maximum of this curve occurs at the point where the slope vanishes

$$\frac{d}{dx}\left(ax - bx^2\right) = a - 2bx = 0 \qquad \text{; maximum}$$

(c) In case (a), the second derivative is $2b > 0$, and the curve is concave *up* at the minimum. In case (b), the second derivative is $-2b < 0$, which is concave *down*. In case (c), the function grows with positive x, and decreases with negative x, so the origin, where the derivative vanishes, cannot be either a minimum or a maximum. In fact, the second derivative vanishes there, and the curve is *flat* at the origin. In this case, the origin is a *point of inflection*.

Problem 1.11 Consider the points $(a, 0)$ and $(0, b)$ on the axes in the (x, y)-plane. The straight line connecting them is $y = b(1 - x/a)$. Show the *area* under this curve is given by

$$\int_0^a y \, dx = \frac{1}{2}ab \qquad \text{; area}$$

Interpret this result in terms of the area of a right-triangle.

Solution to Problem 1.11

The line $y = b(1 - x/a)$ has negative slope $dy/dx = -b/a$. At $x = 0$, it starts at the point $y = b$, and it then comes down to $y = 0$ at $x = a$. The area under this part of the curve is

$$A = b \int_0^a \left(1 - \frac{x}{a} \right) dx = b \left[x - \frac{x^2}{2a} \right]_0^a = b \frac{a}{2}$$

The line and the x- and y-axes define a right triangle of height b and base a. The area is indeed (base $\times$ height)$/2$.

Problem 1.12 (a) A simple example of a *multiple integral*, where the limits are independent and the integrals merely factor is the following

$$\int_0^a dx \int_0^b dy\, x^3 y^4 = \frac{a^4 b^5}{20}$$

Verify this result;

(b) Show

$$\int_0^a dx \int_0^x dy\, x^3 y^4 = \frac{a^9}{45}$$

Solution to Problem 1.12

(a) Here the two integrals *factor*, and

$$\int_0^a dx \int_0^b dy\, x^3 y^4 = \left(\int_0^a x^3\, dx \right) \left(\int_0^b y^4\, dy \right) = \left(\frac{a^4}{4} \right) \left(\frac{b^5}{5} \right)$$

$$= \frac{a^4 b^5}{20}$$

(b) Here the two integrals again *factor*, but the y-integral must be done first since its upper limit depends on x

$$\int_0^a dx \int_0^x dy\, x^3 y^4 = \int_0^a dx\, x^3 \times \left(\int_0^x y^4\, dy \right) = \int_0^a dx\, x^3 \times \left(\frac{x^5}{5} \right)$$

$$= \frac{a^9}{45}$$

Chapter 2

Vectors

Problem 2.1 (a) There is another way of writing the vector product of two vectors. Introduce the completely antisymmetric Levi-Civita tensor[1]

$$\varepsilon_{ijk} = +1 \qquad ; \ (i,j,k) \ \text{an even permutation of} \ (1,2,3)$$
$$= -1 \qquad ; \ (i,j,k) \ \text{an odd permutation of} \ (1,2,3)$$
$$= 0 \qquad \ \ ; \ \text{otherwise}$$

together with the summation convention that repeated Latin indices are summed from 1 to 3. Label the (x, y, z)-axes with $(1, 2, 3)$. Show that

$$(\vec{a} \times \vec{b})_i = \varepsilon_{ijk} a_j b_k$$

(b) Show

$$\varepsilon_{ijk} \varepsilon_{ilm} = \delta_{jl} \delta_{km} - \delta_{jm} \delta_{kl}$$

where δ_{ij} is the Kronecker delta

$$\delta_{ij} = 1 \qquad ; \ \text{if} \ i = j$$
$$= 0 \qquad ; \ \text{if} \ i \neq j$$

Solution to Problem 2.1

(a) There are six non-zero elements to the Levi-Civita tensor

$$\varepsilon_{123} = \varepsilon_{312} = \varepsilon_{231} = +1$$
$$\varepsilon_{213} = \varepsilon_{321} = \varepsilon_{132} = -1$$

---

[1] Here (i, j, k) are *indices*.

With the summation convention, one has

$$\vec{a} \cdot \vec{b} = a_i b_i = a_1 b_1 + a_2 b_2 + a_3 b_3$$

The first component of the indicated cross-product is then given by

$$(\vec{a} \times \vec{b})_1 = \varepsilon_{1jk} a_j b_k = \varepsilon_{123} a_2 b_3 + \varepsilon_{132} a_3 b_2$$
$$= a_2 b_3 - a_3 b_2$$

and similarly for the others. In cartesian notation, this reproduces the expression in Eq. (2.10)

$$\vec{a} \times \vec{b} \equiv (a_y b_z - a_z b_y)\,\hat{x} + (a_z b_x - a_x b_z)\,\hat{y} + (a_x b_y - a_y b_x)\,\hat{z}$$

$$; \text{ vector product}$$

(b) Write out

$$\varepsilon_{ijk}\varepsilon_{ilm} = \varepsilon_{1jk}\varepsilon_{1lm} + \varepsilon_{2jk}\varepsilon_{2lm} + \varepsilon_{3jk}\varepsilon_{3lm}$$

Take any pair of values for (j, k), say $(2, 3)$; they must be different. In this case, only the first term on the r.h.s. $\varepsilon_{1jk}\varepsilon_{1lm}$ contributes. The second pair of indices (l, m) is then also, by necessity, confined to $(2, 3)$; they must also be different. Suppose $j = l$ and $k = m$, the contribution of this term is

$$\varepsilon_{123}\,\varepsilon_{123} = \varepsilon_{132}\,\varepsilon_{132} = +1$$

Suppose $j = m$ and $k = l$, the contribution of this term is

$$\varepsilon_{123}\,\varepsilon_{132} = \varepsilon_{132}\,\varepsilon_{123} = -1$$

Now within this $(2, 3)$-subspace, there are a total of $2^4 = 16$ terms for $\varepsilon_{1jk}\varepsilon_{1lm}$, and only the above four are non-zero. The following expression reproduces these results

$$\delta_{jl}\delta_{km} - \delta_{jm}\delta_{kl} = 0 \qquad\quad ; j = k,\ \underline{\text{or}},\ l = m$$
$$= +1 \qquad\quad ; j = l,\ k = m$$
$$= -1 \qquad\quad ; j = m,\ k = l$$

The other two subspaces, $(1, 2)$ and $(3, 1)$, are handled by the additional two terms on the r.h.s. of the first expression above in part (b).[2] Hence

$$\varepsilon_{ijk}\varepsilon_{ilm} = \delta_{jl}\delta_{km} - \delta_{jm}\delta_{kl}$$

---

[2]This encompasses all the non-zero terms in $\delta_{jl}\delta_{km} - \delta_{jm}\delta_{kl}$. If you are unconvinced, just write out the $3^4 = 81$ possibilities for (j, k, l, m) in $\varepsilon_{ijk}\varepsilon_{ilm} = \delta_{jl}\delta_{km} - \delta_{jm}\delta_{kl}$.

Problem 2.2 Use the results in Prob. 2.1 to prove the vector identity

$$(\vec{a} \times \vec{b}) \cdot (\vec{c} \times \vec{d}) = (\vec{a} \cdot \vec{c})(\vec{b} \cdot \vec{d}) - (\vec{a} \cdot \vec{d})(\vec{b} \cdot \vec{c})$$

Solution to Problem 2.2

Use the results in Prob. 2.1 to write

$$\begin{aligned}
(\vec{a} \times \vec{b}) \cdot (\vec{c} \times \vec{d}) &= \varepsilon_{ijk}\,\varepsilon_{ilm}\,a_j b_k c_l d_m \\
&= [\delta_{jl}\delta_{km} - \delta_{jm}\delta_{kl}]\,a_j b_k c_l d_m \\
&= (\vec{a} \cdot \vec{c})(\vec{b} \cdot \vec{d}) - (\vec{a} \cdot \vec{d})(\vec{b} \cdot \vec{c})
\end{aligned}$$

Problem 2.3 Show that the vector triple product is invariant under cyclic permutations

$$\vec{a} \cdot (\vec{b} \times \vec{c}) = \vec{c} \cdot (\vec{a} \times \vec{b}) = \vec{b} \cdot (\vec{c} \times \vec{a})$$

Solution to Problem 2.3

It is readily established that the Levi-Civita tensor is invariant under a cyclic permutation of its indices

$$\varepsilon_{ijk} = \varepsilon_{kij} = \varepsilon_{jki}$$

Hence

$$\begin{aligned}
\vec{a} \cdot (\vec{b} \times \vec{c}) &= \varepsilon_{ijk} a_i b_j c_k \\
&= \varepsilon_{kij} c_k a_i b_j = \varepsilon_{jki} b_j c_k a_i
\end{aligned}$$

It follows that

$$\vec{a} \cdot (\vec{b} \times \vec{c}) = \vec{c} \cdot (\vec{a} \times \vec{b}) = \vec{b} \cdot (\vec{c} \times \vec{a})$$

Problem 2.4 Show that

$$\vec{a} \times (\vec{b} \times \vec{c}) = (\vec{a} \cdot \vec{c})\,\vec{b} - (\vec{a} \cdot \vec{b})\,\vec{c}$$

Solution to Problem 2.4

In a similar fashion, use the result in Problem 2.1 to write

$$\left[\vec{a} \times (\vec{b} \times \vec{c})\right]_j = \varepsilon_{jki}\, a_k \left(\varepsilon_{ilm}\, b_l c_m\right) = \varepsilon_{ijk}\, \varepsilon_{ilm}\, a_k b_l c_m$$
$$= \left[\delta_{jl}\delta_{km} - \delta_{jm}\delta_{kl}\right] a_k b_l c_m$$
$$= (\vec{a} \cdot \vec{c})\, b_j - (\vec{a} \cdot \vec{b})\, c_j$$

In vector form this reads

$$\vec{a} \times (\vec{b} \times \vec{c}) = (\vec{a} \cdot \vec{c})\, \vec{b} - (\vec{a} \cdot \vec{b})\, \vec{c}$$

Problem 2.5 There are two operators in vector calculus that are used all the time in analysis, the divergence and the curl.[3]

(a) The divergence of a vector field $\vec{v}(\vec{x}, t)$ is defined by

$$\text{div } \vec{v}(\vec{x}, t) \equiv \vec{\nabla} \cdot \vec{v}(\vec{x}, t)$$

Show

$$\vec{\nabla} \cdot \vec{v}(\vec{x}, t) = \frac{\partial v_x(\vec{x}, t)}{\partial x} + \frac{\partial v_y(\vec{x}, t)}{\partial y} + \frac{\partial v_z(\vec{x}, t)}{\partial z} \qquad ; \text{ divergence}$$

(b) The curl of a vector field is defined by

$$\text{curl } \vec{v}(\vec{x}, t) \equiv \vec{\nabla} \times \vec{v}(\vec{x}, t)$$

Show

$$\vec{\nabla} \times \vec{v}(\vec{x}, t) = \left(\frac{\partial v_z}{\partial y} - \frac{\partial v_y}{\partial z}\right) \hat{x} + \left(\frac{\partial v_x}{\partial z} - \frac{\partial v_z}{\partial x}\right) \hat{y} + \left(\frac{\partial v_y}{\partial x} - \frac{\partial v_x}{\partial y}\right) \hat{z}$$
$$; \text{ curl}$$

Solution to Problem 2.5

The gradient is defined by

$$\vec{\nabla} \equiv \hat{x}\frac{\partial}{\partial x} + \hat{y}\frac{\partial}{\partial y} + \hat{z}\frac{\partial}{\partial z} \qquad ; \text{ gradient}$$

where the cartesian basis vectors are orthonormal and constant, and the partial derivatives keep the other variables in the set (x, y, z) fixed.

---

[3]We leave two crucial theorems in vector calculus, Gauss' theorem on the integral over the divergence, and Stokes' theorem on the integral over the curl, for the next course *Introduction to Electricity and Magnetism* [Walecka (2018)], where they play a central role.

A vector field has the form

$$\vec{v}(\vec{x}, t) = v_x(\vec{x}, t)\,\hat{x} + v_y(\vec{x}, t)\,\hat{y} + v_z(\vec{x}, t)\,\hat{z} \qquad ; \text{ vector field}$$

(a) The divergence of the vector field is then given by

$$\vec{\nabla} \cdot \vec{v}(\vec{x}, t) = \frac{\partial v_x(\vec{x}, t)}{\partial x} + \frac{\partial v_y(\vec{x}, t)}{\partial y} + \frac{\partial v_z(\vec{x}, t)}{\partial z} \qquad ; \text{ divergence}$$

(b) The curl of the vector field is given by the following determinant

$$\vec{\nabla} \times \vec{v}(\vec{x}, t) = \det \begin{vmatrix} \hat{x} & \hat{y} & \hat{z} \\ \partial/\partial x & \partial/\partial y & \partial/\partial z \\ v_x & v_y & v_z \end{vmatrix}$$

Thus

$$\vec{\nabla} \times \vec{v}(\vec{x}, t) = \left(\frac{\partial v_z}{\partial y} - \frac{\partial v_y}{\partial z} \right)\hat{x} + \left(\frac{\partial v_x}{\partial z} - \frac{\partial v_z}{\partial x} \right)\hat{y} + \left(\frac{\partial v_y}{\partial x} - \frac{\partial v_x}{\partial y} \right)\hat{z}$$

$$; \text{ curl}$$

Problem 2.6 Show the gradient yields the following relation[4]

$$-\vec{\nabla}\frac{1}{|\vec{r} - \vec{r}_2|} = \frac{(\vec{r} - \vec{r}_2)}{|\vec{r} - \vec{r}_2|^3}$$

Solution to Problem 2.6

Let us do one component

$$-\hat{x}\frac{\partial}{\partial x}\left[(x - x_2)^2 + (y - y_2)^2 + (z - z_2)^2\right]^{-1/2} =$$

$$\frac{(x - x_2)\hat{x}}{\left[(x - x_2)^2 + (y - y_2)^2 + (z - z_2)^2\right]^{3/2}}$$

Since this holds for all three components

$$-\vec{\nabla}\frac{1}{|\vec{r} - \vec{r}_2|} = \frac{(\vec{r} - \vec{r}_2)}{|\vec{r} - \vec{r}_2|^3}$$

Problem 2.7 Use the definition of the dot product and the chain rule to show

$$\frac{d}{dt}\left[\frac{1}{2}\left(\frac{d\vec{r}}{dt}\right)^2\right] = \left(\frac{d\vec{r}}{dt}\right) \cdot \left(\frac{d^2\vec{r}}{dt^2}\right)$$

---

[4] *Hint*: Write $|\vec{r} - \vec{r}_2|^{-1} = \left[(x - x_2)^2 + (y - y_2)^2 + (z - z_2)^2\right]^{-1/2}$.

This relation is used in the discussion of energy in Sec. 6.3.2.

Solution to Problem 2.7

This relation, in detail, is

$$
\frac{1}{2}\frac{d}{dt}\left[\left(\frac{d\vec{r}}{dt}\right)\cdot\left(\frac{d\vec{r}}{dt}\right)\right] = \frac{1}{2}\frac{d}{dt}\left[\left(\frac{dx}{dt}\right)^2 + \left(\frac{dy}{dt}\right)^2 + \left(\frac{dz}{dt}\right)^2\right]
$$

$$
= \left(\frac{dx}{dt}\right)\left(\frac{d^2x}{dt^2}\right) + \left(\frac{dy}{dt}\right)\left(\frac{d^2y}{dt^2}\right) + \left(\frac{dz}{dt}\right)\left(\frac{d^2z}{dt^2}\right)
$$

$$
= \left(\frac{d\vec{r}}{dt}\right)\cdot\left(\frac{d^2\vec{r}}{dt^2}\right)
$$

Chapter 3

Inertial Coordinate Systems

Problem 3.1 Suppose one sits in a coordinate system that is undergoing uniform *acceleration* with respect to the fixed stars so that (see Fig. 3.1 in the text)

$$\vec{a}_R = \frac{d\vec{v}_R}{dt} = \text{constant}$$

Show that in this frame, Newton's second law is modified by the addition of an *inertial force*

$$\vec{F}_{\text{inertial}} = -m\vec{a}_R \qquad ; \text{ inertial force}$$

Solution to Problem 3.1

From Eq. (3.2), the velocity $\vec{v}_0$ in the primary inertial frame is related to the velocity $\vec{v}$ observed in a frame that is moving with a velocity $\vec{v}_R$ with respect to the fixed stars as given by

$$\vec{v}_0 = \vec{v}_R + \vec{v} \qquad ; \text{ in primary frame}$$

Suppose the moving frame is actually *accelerating* in the primary inertial frame so that

$$\vec{a}_R = \frac{d\vec{v}_R}{dt} = \text{constant}$$

Newton's second law is formulated in the primary inertial coordinate system. If a particle of mass m is acted upon by force $\vec{F}$, then Newton's second law says

$$m\frac{d\vec{v}_0}{dt} = \vec{F} \qquad ; \text{ Newton's law}$$

17

It follows that

$$m\frac{d\vec{v}}{dt} = \vec{F} + \vec{F}_{\text{inertial}}$$

where the inertial force acting in the accelerating coordinate system is given by

$$\vec{F}_{\text{inertial}} = -m\vec{a}_R \qquad ; \text{ inertial force}$$

Chapter 4

Newton's Laws

Problem 4.1 (a) Start in the primary inertial coordinate system and consider the gravitational force between two particles in Eq. (4.5). Show that this force *remains the same in all inertial frames*.

(b) Repeat for any two-body force that only depends on the interparticle separation $\vec{r} = \vec{r}_1 - \vec{r}_2$.

Solution to Problem 4.1

Newton's law of gravitation in Eq. (4.5) reads

$$\vec{F}_{21} = -Gm_1 m_2 \frac{(\vec{r}_1 - \vec{r}_2)}{|\vec{r}_1 - \vec{r}_2|^3} \quad ; \text{ Newton's gravity}$$

Let a position vector in the primary inertial coordinate system be denoted by $\vec{r}_0$, and consider any other inertial coordinate system whose origin is located at $\vec{v}_R t$, with constant $\vec{v}_R$. The relation between the position vector $\vec{r}$ seen in the frame moving with velocity $\vec{v}_R$ relative to the primary inertial coordinate system, and the position vector $\vec{r}_0$ seen in the primary frame is then[1]

$$\vec{r}_0 = \vec{v}_R t + \vec{r}$$

where it is assumed that the origins coincide at time $t = 0$. The first term on the r.h.s. *cancels in the instantaneous position differences* $\vec{r}_1 - \vec{r}_2$ appearing in the force in Eq. (4.5).

(b) This clearly holds for *any* force depending on $\vec{r}_1 - \vec{r}_2$. The instantaneous position *differences* are the same in any inertial frame.

Problem 4.2 Energy conservation provides a *first integral* of Newton's

---

[1]Differentiation with respect to time reproduces $\vec{v}_0 = \vec{v}_R + \vec{v}$.

second law for the spring, in that given the energy and position, one can determine the particle's velocity, and *vice versa*. Suppose the system has energy E and the particle is at a position $x(t)$, what is the particle's velocity $dx(t)/dt$?

Solution to Problem 4.2

Energy conservation for a particle of mass m connected to a spring with spring constant κ moving in one dimension is

$$E = \frac{1}{2}m\left[\frac{dx(t)}{dt}\right]^2 + \frac{1}{2}\kappa\,[x(t)]^2 = \text{constant}$$

Hence the particle's velocity is given in terms of its energy and position by[2]

$$\frac{dx(t)}{dt} = \pm\left(\frac{1}{m}\left\{2E - \kappa\,[x(t)]^2\right\}\right)^{1/2}$$

Problem 4.3 Suppose one has a particle in the primary inertial frame and there is a local gravitational field which exerts a force $m_G\vec{g}$ on the particle. Now view that situation from another frame undergoing an *acceleration* $\vec{a}_R$ with respect to the primary frame. It follows from Prob. 3.1 that in this accelerating frame, the net effective force on the particle is

$$\vec{F} = m_G\vec{g} - m_I\vec{a}_R$$

(a) Use the equivalence principle to show

$$\vec{F} = m(\vec{g} - \vec{a}_R)$$

(b) Show that if $\vec{a}_R = \vec{g}$, that is, if the frame is *freely falling* in the gravitational field, then the net force vanishes

$$\vec{F} = 0 \qquad ; \text{ if } \vec{a}_R = \vec{g}$$

and the particle feels no force in this frame. The existence of this *local freely falling frame* in which an inertial force exactly cancels the gravitational force is one of the basic tenants of general relativity.

Solution to Problem 4.3

It is shown in Prob. 3.1 that if one is in a coordinate system undergoing an acceleration $\vec{a}_R = d\vec{v}_R/dt$ with respect to the fixed stars, then Newton's

---

[2]The sign is determined by the configuration under examination.

second law reads

$$m_I \frac{d\vec{v}}{dt} = \vec{F}_{\text{applied}} + \vec{F}_{\text{inertial}}$$

$$\vec{F}_{\text{inertial}} = -m_I \vec{a}_R \qquad ; \text{ inertial force}$$

where m_I is the inertial mass. Suppose there is also a local gravitational field $\vec{g}$ acting on the particle. Then

$$m_I \frac{d\vec{v}}{dt} = m_G \vec{g} - m_I \vec{a}_R$$

where m_G is the gravitational mass.

(a) The equivalence principle states that

$$m_G = m_I \equiv m \qquad ; \text{ equivalence principle}$$

Hence, the particle's motion is governed by the relation

$$\frac{d\vec{v}}{dt} = \vec{g} - \vec{a}_R$$

(b) If the accelerating coordinate system is simply freely falling in the gravitational field so that $\vec{a}_R = \vec{g}$, then the force in this equation *vanishes*

$$\frac{d\vec{v}}{dt} = 0 \qquad ; \text{ local freely falling frame}$$

The existence of this *local freely falling frame* in which an inertial force exactly cancels the gravitational force is one of the basic tenants of general relativity.

Problem 4.4 If a particle is sliding on a table in the lab, there is generally a force of *friction* opposing this motion. This force is proportional to the normal force, with a constant of proportionality called the *coefficient of friction* μ_f,[3]

$$F_{\text{friction}} = \mu_f F_{\perp} \qquad ; \text{ friction}$$

Show the equation of motion for the particle sliding in the x-direction on the table top under an applied force F_{app}, *with* friction, is

$$m \frac{d^2 x}{dt^2} = F_{\text{app}} - \mu_f mg \qquad ; \text{ with friction}$$

---

[3] The coefficient of *static* friction is generally larger than the coefficient of *sliding* friction.

It is assumed here that $F_{\text{app}} \geq \mu_f mg$.

Solution to Problem 4.4

If the particle is sliding on a horizontal table here in the lab, then there is a normal force of gravity $F_\perp = mg$ acting on it and pointing down (which is opposed by the upward force the table exerts on the particle). If the particle is acted upon by a horizontal force F_{app} that causes it to move in the x-direction, then Newton's second law reads

$$m\frac{d^2x}{dt^2} = F_{\text{app}} - \mu_f mg \qquad \text{; with friction}$$

where μ_f is the *coefficient of sliding friction*.

Problem 4.5 (a) Consider an object of mass M moving in one dimension. Suppose one strikes the object with a force F over a short time interval Δt, hitting a pool ball with a cue, for example. The *impulse* given to the object is defined by

$$\mathcal{J} \equiv F\Delta t \qquad \text{; impulse}$$

Assume the force is such that the impulse remains finite as $\Delta t \to 0$. Show that it follows from Newton's second law that if the object starts from rest, its momentum after it is struck is

$$Mv = \mathcal{J}$$

Solution to Problem 4.5

Newton's second law in one dimension is the limit of the expression

$$M\frac{\Delta v}{\Delta t} = F$$

Multiply this by Δt to obtain the *impulse*

$$\mathcal{J} \equiv F\Delta t \qquad \text{; impulse}$$

Now assume the force is such that the impulse remains finite as $\Delta t \to 0$. If the object starts from rest so that $\Delta v = v$, its momentum after it is struck is finite and equal to

$$Mv = \mathcal{J}$$

This is what happens when you strike a pool ball in the center with a pool cue.

Chapter 5

Examples

Problem 5.1 A particle of mass m on the earth's surface slides *without* friction down an incline plane whose normal makes an angle α with the vertical direction. Let ζ be the distance down the plane. Use Newton's second law to show that the equation of motion is

$$m\frac{d^2\zeta}{dt^2} = mg\sin\alpha$$

What is the normal (constraint) force exerted by the plane on the particle?

Solution to Problem 5.1

The gravitational force $\vec{F}_{\text{grav}} = m\vec{g}$ points down. Its tangential component down the incline plane and normal component perpendicular to the incline plane are

$$\left(\vec{F}_{\text{grav}}\right)_{\parallel} = mg\sin\alpha \qquad ; \text{ down incline plane}$$

$$\left(\vec{F}_{\text{grav}}\right)_{\perp} = mg\cos\alpha \qquad ; \text{ normal to incline plane}$$

Newton's second law written in a direction down the plane, normal to the frictionless constraint force, reads

$$m\frac{d^2\zeta}{dt^2} = mg\sin\alpha$$

Since there is no motion in a direction normal to the incline plane, the normal force the incline plane exerts on the particle just balances the normal gravitational force the particle exerts on the incline plane

$$F_{\perp} = mg\cos\alpha \qquad ; \text{ incline plane on particle}$$

Problem 5.2 A particle of mass m on the earth's surface slides without friction down a hoop of radius a with a diameter oriented in the vertical direction. Let θ be the angle of the particle with respect to the vertical measured from the center of the hoop. Use Newton's second law to show that the equation of motion is the inverted pendulum equation

$$ma\frac{d^2\theta}{dt^2} = mg\sin\theta$$

Show that for small displacements from the vertical the motion is unstable. Describe that motion.[1]

Solution to Problem 5.2

Introduce polar coordinates for the bead on the hoop of radius a as in Fig. 11.1 in the text.[2] Let $\hat{z}$ be a unit vector in the up z-direction, and $(\hat{r}, \hat{\theta})$ be orthogonal unit vectors located at (r, θ) and pointing in the direction of the increasing coordinates (see Prob. 7.7). A little effort then shows that these basis vectors are related by

$$-\hat{z} = -\hat{r}\,\cos\theta + \hat{\theta}\,\sin\theta$$

The gravitational force on the particle is then given by

$$\vec{F}_{\text{grav}} = -mg\,\hat{z} = -mg\cos\theta\,\hat{r} + mg\sin\theta\,\hat{\theta}$$

The first term on the r.h.s. is balanced by a constraint force that prohibits the bead from moving in the radial direction. The last term gives rise to a tangential acceleration along the hoop, so that Newton's second law reads

$$m\left(\frac{d\vec{v}}{dt}\right)_{\text{tang}} = m\left[\frac{d}{dt}(a\dot{\theta})\right]\hat{\theta} = mg\sin\theta\,\hat{\theta}$$

Hence

$$\ddot{\theta} = \frac{g}{a}\sin\theta \qquad ; \text{ inverted pendulum equation}$$

This is the *inverted pendulum equation.*
For small θ, this equation reads

$$\ddot{\theta} = \omega^2\theta \qquad ; \omega^2 = \frac{g}{a}$$

---

[1] See also Prob. 7.6.
[2] Note that here the hoop is stationary.

The solutions to this equation are $\theta(t) = e^{\pm \omega t}$. Suppose we construct the solution that describes the particle at rest a short distance $\theta(0) = \eta$ from the top of the hoop

$$\theta(t) = \frac{\eta}{2}\left(e^{\omega t} + e^{-\omega t}\right) \qquad ; \; \theta(0) = \eta$$

$$; \; [d\theta(t)/dt]_{t=0} = 0$$

The run-away solution dominates, corresponding to the acceleration of the mass down the hoop.

Problem 5.3 Suppose one includes a *viscous damping force* proportional to the velocity in the example of a dropped object in Sec. 1.3. Newton's second law then reads

$$\frac{dv_z}{dt} = g - \eta v_z$$

where η is a coefficient of viscosity (effectively, of "friction"). Write this as the inhomogeneous differential equation

$$\frac{dv_z}{dt} + \eta v_z = g$$

(a) Show the following provides a solution to this equation

$$v_z(t) = \frac{g}{\eta}\left[1 - e^{-\eta t}\right]$$

(b) Show that for short times $\eta t \ll 1$, this reduces to the previously studied solution

$$v_z(t) = gt \qquad ; \; \eta t \ll 1$$

(c) Show that for long times $\eta t \gg 1$, the particle reaches a *terminal velocity*

$$v_z(t) = \frac{g}{\eta} \qquad ; \; \eta t \gg 1$$

Solution to Problem 5.3

(a) Take a time derivative of the proposed solution

$$\frac{dv_z(t)}{dt} = \frac{g}{\eta}\left[\eta e^{-\eta t}\right]$$

Hence

$$\frac{dv_z}{dt} + \eta v_z = ge^{-\eta t} + g\left[1 - e^{-\eta t}\right]$$
$$= g$$

(b) For short times, $e^{-\eta t} \approx 1 - \eta t$, and therefore

$$v_z(t) = gt \qquad ; \; \eta t \ll 1$$

This is just the result for uniform acceleration in the gravitational field.

(c) For long times $e^{-\eta t} \approx 0$, and

$$v_z(t) = \frac{g}{\eta} \qquad ; \; \eta t \gg 1$$

Thus the friction of the air resistance causes the falling particle to reach a *terminal velocity*.

Chapter 6

Energy

Problem 6.1 Substitute the solutions for the projectile motion in Sec. 5.2 $[x(t), z(t), v_x(t), v_z(t)]$ into the expressions for the energy in Eqs. (6.9), and derive Eq. (6.10) relating the projectile energy to the initial muzzle velocity.

Solution to Problem 6.1

Energy conservation for the particle in the projectile motion analyzed in Sec. 5.2 states that[1]

$$E = T + V = \text{constant}$$
$$T = \frac{m}{2}\left(v_x^2 + v_z^2\right)$$
$$V(z) = mgz$$

Integration of Newton's second law in Sec. 5.2 gives as the solution for $[v_x(t), v_z(t), z(t)]$ for that projectile motion

$$v_x(t) = v_{0x}$$
$$v_z(t) = v_{0z} - gt$$
$$z(t) = v_{0z}t - \frac{1}{2}gt^2$$

Substitution of this solution in the energy gives

$$E = \frac{m}{2}\left[(v_{0x})^2 + (v_{0z} - gt)^2\right] + mg\left(v_{0z}t - \frac{1}{2}gt^2\right)$$

Energy conservation therefore again provides a first integral of Newton's

---

[1] Remember that now z is *up*.

second law, since

$$E = \frac{m}{2}\left[(v_{0x})^2 + (v_{0z})^2\right] = \frac{m}{2}v_0^2$$

which is just the initial kinetic energy of the projectile as it leaves the muzzle at the origin.

Problem 6.2 The force obtained from the gravitational potential $V_G(r)$ in Eq. (6.12) follows from Eq. (2.28) as

$$\vec{F}_G = -\vec{\nabla}V_G(r) = -\frac{dV_G(r)}{dr}\,\hat{r}$$

The work this force does as the second particle moves in from infinity is given by Eq. (6.6) as

$$W = -\int_\infty^r d\vec{r} \cdot \vec{F}_G \qquad ; \; d\vec{r} = \hat{r}\,dr$$

Show

$$W = V_G(r)$$

Thus the work done by the gravitational field when the particle is moved in from infinity is just the (negative) gravitational potential energy at the radius r.[2]

Solution to Problem 6.2

Given particle 1 of mass m_1 located at $\vec{r}_1$ and particle 2 of mass m_2 located at $\vec{r}_2$, the gravitational potential energy of the pair in Eq. (6.12) is

$$V_G(r) = -G\frac{m_1 m_2}{r} \qquad ; \; \vec{r} \equiv \vec{r}_1 - \vec{r}_2$$

where $r = |\vec{r}|$. The gravitational force exerted by the second particle on the first is

$$\vec{F}_{21}(\vec{r}) = -(Gm_1 m_2)\frac{\vec{r}}{r^3} = -\vec{\nabla}V_G(r)$$

From Eq. (2.28)

$$\vec{\nabla}V_G(r) = \frac{dV_G(r)}{dr}\,\hat{r}$$

---

[2]A proper determination of the *sign* of the work done here is given in the detailed solution to Prob. 6.2 below (we must be careful with the sign of dr).

Now fix the second particle and let me move the first particle from the position r out to ∞. If I hold on to the first particle, I must exert a force just slightly larger than $-\vec{F}_{21}$ to move it out quasistatically, and hence the work that I do against the gravitational force is

$$dW = -\vec{F}_{21} \cdot d\vec{r} = \frac{dV_G(r)}{dr} dr$$

The total positive work I do on the system to move the first mass out to infinity is

$$W = \int_r^\infty \frac{dV_G(r)}{dr} dr = [V_G(r)]_r^\infty = -V_G(r)$$

As a check on signs we can employ the first law of thermodynamics (with no heat flow).[3] The initial and final energies of the system are

$$E_i = V_G(r) \qquad ; E_f = V_G(\infty) = 0$$

The increase in energy of the system is then

$$\Delta E = E_f - E_i = -V_G(r)$$

The work done *on the system* is calculated above to be $\Delta W = -V_G(r)$. The first law of thermodynamics then says that

$$\Delta E = \Delta W \qquad ; \text{first law of thermodynamics}$$
$$= -V_G(r)$$

Problem 6.3 The gravitational potential energy of two bodies located at $\vec{r}_1$ and $\vec{r}_2$ follows from Eqs. (6.12) as

$$V_G = -G \frac{m_1 m_2}{|\vec{r}_1 - \vec{r}_2|}$$

Define the corresponding *gravitational potential* at the position $\vec{r} \equiv \vec{r}_1$ by

$$U(\vec{r}) \equiv -G \frac{m_2}{|\vec{r} - \vec{r}_2|} \qquad ; \text{gravitational potential}$$

It follows from Eqs. (6.12) that the gravitational force on a particle with mass μ located at $\vec{r}$ is then given by[4]

$$\vec{F} = -\mu \vec{\nabla} U(\vec{r}) \qquad ; \text{gravitational force}$$

---

[3]See Chapter 16.
[4]See also Prob. 2.6.

(a) With several bodies surrounding the first one, the gravitational potential becomes

$$U(\vec{r}) = -G \sum_{j=1}^{N} \frac{m_j}{|\vec{r} - \vec{r}_j|}$$

Assume the m_j are identical, with $m_j \equiv m$. Divide space up into many cells of volume $d^3 x_j$, and let $\vec{r}_j$ point to the center of the jth cell. Introduce the *mass density* $\rho(\vec{r}_j)$ where

$$\rho(\vec{r}_j) \equiv m \times (\text{number of particles in the } jth \text{ cell})$$

Show that the gravitational potential from the particles in the tiny volume $d^3 x_j$ then takes the form

$$U(\vec{r}) \approx -G \sum_{j=1}^{N} \frac{\rho(\vec{r}_j) d^3 x_j}{|\vec{r} - \vec{r}_j|}$$

(b) Show that in the limit that the mass density becomes a continuous function, this becomes the integral

$$U(\vec{r}) = -G \int \frac{\rho(\vec{r}') d^3 x'}{|\vec{r} - \vec{r}'|} \qquad ; \text{ gravitational potential}$$

This is a very important result, since it allows us to compute the gravitational potential arising from an arbitrary mass distribution.

Solution to Problem 6.3

Suppose one has particle 1 of mass μ located at a position $\vec{r}$, and a second particle of mass m_2 located at $\vec{r}_2$. The *gravitational potential* at $\vec{r}$ created by the second particle is defined by

$$U(\vec{r}) \equiv -G \frac{m_2}{|\vec{r} - \vec{r}_2|} \qquad ; \text{ gravitational potential}$$

Use the following result from Prob. 2.6

$$-\vec{\nabla} \frac{1}{|\vec{r} - \vec{r}_2|} = \frac{(\vec{r} - \vec{r}_2)}{|\vec{r} - \vec{r}_2|^3}$$

The force on the first particle due to the second is then

$$\vec{F}_{21} = -\mu G \frac{m_2 (\vec{r} - \vec{r}_2)}{|\vec{r} - \vec{r}_2|^3}$$

$$= -\mu \vec{\nabla} U(\vec{r})$$

If we know the gravitational potential $U(\vec{r})$ at $\vec{r}$, then we can compute the gravitational force at $\vec{r}$.

(a) With several bodies surrounding the first one, the gravitational potential is extended to read

$$U(\vec{r}) = -G\sum_{j=1}^{N} \frac{m_j}{|\vec{r} - \vec{r}_j|}$$

Assume the m_j are identical, with $m_j \equiv m$. Divide space up into many cells of volume d^3x_j, and let $\vec{r}_j$ point to the center of the *jth* cell.[5] Introduce the *mass density* $\rho(\vec{r}_j)$ where

$$\rho(\vec{r}_j) \equiv m \times \text{(number of particles in the } jth \text{ cell)}$$

The gravitational potential from the particles in the tiny volumes d^3x_j then takes the approximate form

$$U(\vec{r}) \approx -G\sum_{j=1}^{N} \frac{\rho(\vec{r}_j)d^3x_j}{|\vec{r} - \vec{r}_j|}$$

(b) Take the limit of this expression where the differential volume becomes very small, the mass density becomes a continuous function, and the position vector $\vec{r}_j$ in the denominator locating the little volume becomes precisely defined. In this limit, the above becomes the following integral

$$U(\vec{r}) = -G\int \frac{\rho(\vec{r}\,')d^3x'}{|\vec{r} - \vec{r}\,'|} \qquad ; \text{ gravitational potential}$$

As stated in the problem, *this is a very important result, since it allows us to compute the gravitational potential arising from an arbitrary mass distribution, and from this, the gravitational force.*

Problem 6.4 (a) Suppose one is *outside* of a spherically symmetric mass distribution with $r > r'$. Show the denominator in Eq. (17.65) can be expanded as

$$\left[(\vec{r} - \vec{r}\,')^2\right]^{-1/2} = \left[r^2 - 2r\,\hat{r}\cdot\vec{r}\,' + r'^2\right]^{-1/2}$$
$$= \frac{1}{r}\left(1 + \frac{\hat{r}\cdot\vec{r}\,'}{r} + \cdots\right)$$
$$= \frac{1}{r}\left(1 + \frac{r'\cos\theta}{r} + \cdots\right)$$

---

[5]There is a misprint in the problem in the text where "*ith*" appears instead of "*jth*".

where $\hat{r} \cdot \hat{r}' \equiv \cos\theta$.

(b) Show that the gravitational potential at $\vec{r}$ then becomes

$$U(\vec{r}) = -\frac{G}{r} \int \rho(r')d^3x' \left(1 + \frac{r'\cos\theta}{r} + \cdots \right)$$

(c) Make use of the volume element in spherical coordinates to show that for a spherically symmetric mass distribution, the first correction *vanishes*.[6] Therefore, through this order,

$$U(r) = -\frac{GM}{r}$$

where M is the total mass of the object. Thus through this order, *if one is outside of a spherically symmetric mass distribution, it acts like a point mass at the origin!*[7]

Legend has it that Newton actually delayed publication of his famous work for several years until he could establish this result.

Solution to Problem 6.4

(a) Just for fun, we will do *two* terms in the expansion of the denominator in Eq. (17.65) for $r > r'$. Use the result in Prob. 10.3 to obtain

$$\left[(\vec{r} - \vec{r}')^2\right]^{-1/2} = \left[r^2 - 2r\,\hat{r} \cdot \vec{r}' + r'^2\right]^{-1/2}$$

$$= \frac{1}{r}\left\{1 - \frac{1}{2}\left[-\frac{2\hat{r} \cdot \vec{r}'}{r} + \left(\frac{r'}{r}\right)^2\right] + \frac{3}{8}\left[-\frac{2\hat{r} \cdot \vec{r}'}{r}\right]^2 + \cdots \right\}$$

$$= \frac{1}{r}\left\{1 + \left(\frac{r'}{r}\right)\cos\theta + \left(\frac{r'}{r}\right)^2 \frac{1}{2}(3\cos^2\theta - 1) + \cdots \right\}$$

where $\hat{r} \cdot \hat{r}' \equiv \cos\theta$.

---

[6] Recall $d^3x = r^2\sin\theta\,dr\,d\theta\,d\phi$.

[7] This result actually holds to all orders, since the denominator in Eq. (17.65) can be expanded as

$$\frac{1}{|\vec{r} - \vec{r}'|} = \frac{1}{r_>}\sum_{l=0}^{\infty}\left(\frac{r_<}{r_>}\right)^l P_l(\cos\theta)$$

where $P_l(\cos\theta)$ is the *Legendre polynomial*, satisfying

$$\int_0^\pi P_l(\cos\theta)\sin\theta\,d\theta = 2\delta_{l,0}$$

Here $P_0(\cos\theta) = 1$, $P_1(\cos\theta) = \cos\theta$, $P_2(\cos\theta) = (3\cos^2\theta - 1)/2$, etc.

(b) The gravitational potential at $\vec{r}$ then becomes

$$U(\vec{r}) = -\frac{G}{r} \int \rho(r')d^3x' \left[P_0(\cos\theta) + \left(\frac{r'}{r}\right) P_1(\cos\theta) \right.$$
$$\left. + \left(\frac{r'}{r}\right)^2 P_2(\cos\theta) + \cdots \right]$$

Here $P_l(\cos\theta)$ are the Legendre polynomials

$$P_0(\cos\theta) = 1 \quad ; \ P_1(\cos\theta) = \cos\theta \quad ; \ P_2(\cos\theta) = \tfrac{1}{2}(3\cos^2\theta - 1)$$

(c) The volume element in spherical coordinates for the spherically-symmetric mass density over which we are integrating is

$$d^3x' = r'^2 dr' \sin\theta \, d\theta \, d\phi$$

Now make use of the orthogonality relation for the Legendre polynomials in the previous footnote[8]

$$\int_0^\pi P_l(\cos\theta) \sin\theta \, d\theta = 2\delta_{l,0}$$

Therefore, through this order,

$$U(r) = -\frac{GM}{r}$$

where M is the total mass of the object. Thus through this order, *if one is outside of a spherically symmetric mass distribution, it acts like a point mass at the origin!*

Problem 6.5 (a) Use the result in the previous problem to show that the magnitude of the acceleration of gravity at the earth's surface is given by

$$g = \frac{GM_e}{R_e^2} \quad ; \ \text{acceleration of gravity}$$

where M_e is the earth's mass, R_e its radius, and G is Newton's constant in Eq. (4.6).

(b) Compute g using the following values

$$M_e = 5.976 \times 10^{24} \, \text{kg}$$
$$R_e = 6.378 \times 10^6 \, \text{m}$$

Compare with the measured value in Eq. (1.14).

---

[8]You can easily verify this for the three terms above.

Solution to Problem 6.5

(a) From the previous problem, if one is outside of a spherically symmetric body, the gravitational force acts as if it were a point mass at the origin. Thus, at the earth's surface, the magnitude of the gravitational force is

$$mg = \frac{GM_e m}{R_e^2} \qquad ; \text{gravitational force}$$

Hence, the acceleration of gravity measured at the earth's surface is

$$g = \frac{GM_e}{R_e^2} \qquad ; \text{acceleration of gravity}$$

(b) Newton's gravitational constant is from Eq. (4.6)

$$G = 6.673 \times 10^{-11} \frac{\text{m}^3}{\text{kg-s}^2}$$

We can thus put in some numbers for g

$$g = \frac{\left(6.673 \times 10^{-11}\,\text{m}^3/\text{kg-sec}^2\right)\left(5.976 \times 10^{24}\,\text{kg}\right)}{\left(6.378 \times 10^6\,\text{m}\right)^2}$$

$$= 9.80\,\text{m}/\text{sec}^2$$

The measured value in Eq. (1.14) is

$$g = 9.8\,\text{m}/\text{sec}^2 \qquad ; \text{acceleration of gravity}$$

Problem 6.6 (a) Suppose one is *inside* of a spherically symmetric mass shell with $r' > r$. Show the denominator in Eq. (17.65) can now be expanded as

$$\left[(\vec{r} - \vec{r}')^2\right]^{-1/2} = \left[r'^2 - 2r\,\hat{r} \cdot \vec{r}' + r^2\right]^{-1/2}$$

$$= \frac{1}{r'}\left(1 + \frac{r\,\hat{r} \cdot \vec{r}'}{r'^2} + \cdots\right)$$

$$= \frac{1}{r'}\left(1 + \frac{r\cos\theta}{r'} + \cdots\right)$$

where $\hat{r} \cdot \hat{r}' \equiv \cos\theta$.

(b) Show that the gravitational potential at $\vec{r}$ then becomes

$$U(\vec{r}) = -G\int \frac{\rho(r')}{r'} d^3x' \left(1 + \frac{r\cos\theta}{r'} + \cdots\right)$$

(c) Show that for a spherically symmetric mass shell distribution, the first correction again vanishes. Therefore, through this order,

$$U(r) = -\frac{GM}{r'}$$

where M is the total mass of the object;

(d) The above result is independent of r. Show that it follows from Eq. (17.61) that the gravitational force at $\vec{r}$ then vanishes. Thus through this order, *if one is inside of a spherically symmetric mass shell, there is no gravitational force!*[9]

Solution to Problem 6.6

(a) We will again do *two* terms in the expansion of the denominator in Eq. (17.65) for $r < r'$. Use the result in Prob. 10.3 to obtain

$$\left[(\vec{r} - \vec{r}')^2\right]^{-1/2} = \left[r'^2 - 2r'\,\hat{r}' \cdot \vec{r} + r^2\right]^{-1/2}$$

$$= \frac{1}{r'}\left\{1 - \frac{1}{2}\left[-\frac{2\hat{r}' \cdot \vec{r}}{r'} + \left(\frac{r}{r'}\right)^2\right] + \frac{3}{8}\left[-\frac{2\hat{r}' \cdot \vec{r}}{r'}\right]^2 + \cdots\right\}$$

$$= \frac{1}{r'}\left\{1 + \left(\frac{r}{r'}\right)\cos\theta + \left(\frac{r}{r'}\right)^2 \frac{1}{2}\left(3\cos^2\theta - 1\right) + \cdots\right\}$$

where $\hat{r}' \cdot \hat{r} \equiv \cos\theta$.

(b) The gravitational potential at $\vec{r}$ then becomes

$$U(\vec{r}) = -\frac{G}{r'}\int \rho(r')d^3x'\left[P_0(\cos\theta) + \left(\frac{r}{r'}\right)P_1(\cos\theta)\right.$$

$$\left. + \left(\frac{r}{r'}\right)^2 P_2(\cos\theta) + \cdots\right]$$

where $P_l(\cos\theta)$ are again the Legendre polynomials

$$P_0(\cos\theta) = 1 \quad ; \quad P_1(\cos\theta) = \cos\theta \quad ; \quad P_2(\cos\theta) = \tfrac{1}{2}(3\cos^2\theta - 1)$$

(c) The volume element in spherical coordinates for the spherically-symmetric mass shell over which we are integrating is

$$d^3x' = r'^2 dr' \sin\theta\, d\theta\, d\phi$$

---

[9]It follows from the previous footnote that this result again holds to all orders.

We again make use of the orthogonality relation for the Legendre polynomials in the previous footnote[10]

$$\int_0^\pi P_l(\cos\theta)\sin\theta\,d\theta = 2\delta_{l,0}$$

Therefore, through this order,

$$U(\vec{r}) = -\frac{GM}{r'}$$

where M is the total mass of the shell. This expression in independent of $\vec{r}$. *Thus through this order, if one is inside of a spherically symmetric mass shell, there is no gravitational force!*

Problem 6.7 A tunnel is dug straight through the earth from one surface to another. A particle of mass m slides without friction on a rail through that tunnel. Make use of the results of Probs. 6.4–6.6.

(a) The mass density of the earth (assumed uniform) is

$$\rho_e = \frac{M_e}{4\pi R_e^3/3}$$

Show the gravitational force per unit mass at a point in the tunnel at a distance r from the center of the earth is

$$\vec{f}_G(r) = -\frac{G}{r^2}\left(\frac{4\pi r^3 \rho_e}{3}\right)\hat{r}$$

$$= -\frac{GM_e}{R_e^3}r\,\hat{r}$$

Compute the work done in moving a unit-mass particle out from the origin

$$W(r) = -\int_0^r \vec{f}_G(r)\cdot d\vec{r} = \frac{GM_e}{R_e^3}\frac{r^2}{2}$$

To within a constant, this is the gravitational potential[11]

$$U(r) = \frac{GM_e}{R_e^3}\frac{r^2}{2} + U(0)$$

(b) Write $r^2 = l^2 + x^2$ where l is the distance of closest approach to the origin, and include a kinetic energy for the motion of the particle along the

---

[10]You can easily verify this for the three terms above.

[11]We can even find the constant, since we know $U(R) = -GM_e/R_e$. As a check on Eq. (17.77), we have $\vec{f}_G(r) = -\vec{\nabla}U(r)$.

tunnel to write the particle energy is

$$E(\dot{x}, x) = \frac{1}{2}m\dot{x}^2 + \frac{1}{2}m\left(\frac{GM_e}{R_e^3}\right)x^2 + E_0$$

where E_0 is the energy of the particle at rest at l. Identify this result with an equation for one-dimensional simple harmonic motion about the midpoint of the tunnel.

(c) Conclude that if the particle is released from rest at the surface, the time it takes to reach the other side is independent of the location of the tunnel. Calculate this time.

Solution to Problem 6.7

(a) We have shown that the gravitational force per unit mass at a point in the tunnel a distance r from the center of the sperically symmetric earth comes only from the mass inside that radius. Hence

$$\vec{f}_G(r) = -\frac{G}{r^2}\left(\frac{4\pi r^3 \rho_e}{3}\right)\hat{r}$$

$$= -\frac{GM_e}{R_e^3}r\,\hat{r}$$

The work I do in moving a unit-mass particle out from the origin to r is then

$$W(r) = -\int_0^r \vec{f}_G(r) \cdot d\vec{r} = \frac{GM_e}{R_e^3}\frac{r^2}{2}$$

This is the gravitational potential.

$$U(r) = \frac{GM_e}{R_e^3}\frac{r^2}{2} + U(0)$$

Since we know that $U(R_e) = -GM_e/R_e$ the constant $U(0)$ is determined, and

$$U(r) = -\frac{GM_e}{R_e}\left[\frac{3}{2} - \frac{r^2}{2R_e^2}\right]$$

(b) Now write $r^2 = l^2 + x^2$ where l is the distance of closest approach to the origin, and include a kinetic energy for the motion of the particle along the tunnel to write the particle energy as

$$E(\dot{x}, x) = \frac{1}{2}m\dot{x}^2 + \frac{1}{2}m\left(\frac{GM_e}{R_e^3}\right)x^2 + E_0$$

where $E_0 = mU(l)$ is the energy of the particle at rest at l.

The above result is the energy equation for one-dimensional simple harmonic motion of the particle along the tunnel, about its midpoint. The angular frequency of the oscillator is evidently

$$\omega^2 = \frac{GM_e}{R_e^3}$$

(c) The period of the oscillator is $\tau = 2\pi/\omega$. The time for half the period, which is the time it takes to get from one side of the earth to the other, is

$$\tau_{1/2} = \pi \left(\frac{R_e^3}{GM_e} \right)^{1/2}$$

Just as in Prob. 7.4, with the constants in Prob. 6.5, this gives

$$\tau_{1/2} = \pi \left[\frac{(6.378 \times 10^6 \, \text{m})^3}{(6.673 \times 10^{-11} \, \text{m}^3/\text{kg-s}^2) \times (5.976 \times 10^{24} \, \text{kg})} \right]^{1/2}$$

Hence, the transit time is

$$\tau_{1/2} = 2.53 \times 10^3 \, \text{s} = 42.2 \, \text{minutes}$$

Problem 6.8 A bead of mass m slides without friction along a wire. It starts at a height h above the earth's surface. As the bead drops, the wire has a complicated shape, with many twists, turns, and loops. At the end, the bead rises along a straight section of wire. How high does it rise? Why?

Solution to Problem 6.8

This is a very simple problem, but it is also thought-provoking. For a short time, the earth's surface provides an inertial frame. The particle starts off at rest at a distance h above the surface. Its initial energy is all potential $E = V = mgh$. As it falls, its energy receives an increasing kinetic contribution $T = m\vec{v}^2/2$. As the bead drops, the wire has a complicated shape, with many twists, turns, and loops. The detailed analysis of the motion from Newton's second law, with all the constraint forces, is clearly a challenging problem. However, at the end, as the bead rises along a straight section of wire, it clearly slows down, and its energy again becomes all potential. If there is no friction or dissipation, *energy conservation dictates that the height it rises is again given by* $mgh = E$, the same height it started from! A remarkable result.

Problem 6.9 A particle moving in a plane bounces elastically off a transverse wall, which can only exert a normal force on the particle. Show that the angle of reflection, measured with respect to the normal, is equal to the angle of incidence.

Solution to Problem 6.9

The wall exerts no tangential force, and therefore the tangential component of its momentum, or velocity v_{tang}, is unchanged. The wall does no work during the collision, and hence the magnitude of its kinetic energy, or $|\vec{v}|^2$, is unchanged. Hence the sine of the angle of reflection, measured with respect to the normal, is equal to the sine of the angle of incidence

$$\sin \theta_{\text{inc}} = \frac{v_{\text{tang}}}{|\vec{v}|} = \sin \theta_{\text{refl}}$$

Thus the angles themselves are equal.

Problem 6.10 (a) A particle of mass m and velocity v strikes a target particle of mass M at rest in the laboratory, and is captured by it. Use momentum conservation to show that the energy ε^* available to excite the target is

$$\varepsilon^* = \frac{1}{2}mv^2 \left(1 - \frac{m}{m+M} \right)$$

Because of target recoil, this is less than the incident kinetic energy.

(b) Suppose the laboratory frame is also a C-M frame where the target is moving with a velocity $-mv/M$.[12] Show that in this case

$$\varepsilon^* = \frac{1}{2}mv^2 \left(1 + \frac{m}{M} \right)$$

Now all of the initial kinetic energy can go into target excitation.

Solution to Problem 6.10

(a) *Momentum conservation* states that if there are no external forces, then the total momentum of the system is a constant of the motion [see Eqs. (8.4) and (8.7)]. Here, the initial momentum is $p_i = mv$, and the initial kinetic energy is $T_{\text{in}} = mv^2/2$. The final momentum is $p_f = (m+M)v_f$, and by momentum conservation, this must be equal to p_i. The energy available

---

[12]The C-M frame is also the center-of-momentum frame [see Eq. (8.11)].

to excite the target is therefore

$$
\varepsilon^* = \frac{1}{2}mv^2 - \frac{1}{2}(m+M)v_f^2 = \frac{1}{2}mv^2 - \frac{1}{2}(m+M)\left(\frac{mv}{m+M}\right)^2
$$

$$
= \frac{1}{2}mv^2\left(1 - \frac{m}{m+M}\right)
$$

(b) In this case, the final system is at *rest* in the C-M system, and the total initial kinetic energy is available to excite the target

$$
\varepsilon^* = \frac{1}{2}mv^2 + \frac{1}{2}M\left(\frac{mv}{M}\right)^2 = \frac{1}{2}mv^2\left(1 + \frac{m}{M}\right)
$$

Problem 6.11 (a) Show that with the viscous damping term in Prob. 5.3, the energy of the falling particle is *not conserved.*

(b) Where does the kinetic energy go?

(c) Where does the argument in Sec. 6.1 that derives energy conservation from Newton's second law break down?

Solution to Problem 6.11

(a) If the falling particle reaches a terminal velocity (and hence a terminal *kinetic energy*) while its potential energy continues to decrease, then its energy is clearly *not* conserved;

(b) The energy from the viscous damping goes into heating the surrounding air, the source of the damping;

(c) Since the viscous damping term in Newtons's second law is proportional to the velocity dx/dt, one cannot make use of the trick employed in Eqs. (6.1)–(6.2) to derive energy conservation.[13]

---

[13]There $x(dx/dt) = d(x^2/2)/dt$; here $(dx/dt)^2 = d(\cdots)/dt$??

Chapter 7

Angular Momentum

Problem 7.1 Suppose we have motion in the (x, y)-plane, and that motion is viewed from a coordinate system *rotating about the z-axis*. If the second, primed coordinate system has rotated by an angle $\theta(t)$, the relations between the two sets of unit basis vectors is as follows

$$\hat{x}' = \hat{x} \cos\theta + \hat{y} \sin\theta$$
$$\hat{y}' = -\hat{x} \sin\theta + \hat{y} \cos\theta$$

At any instant, a spatial vector $\vec{r}$ can be expressed in either frame as

$$\vec{r} = x\,\hat{x} + y\,\hat{y} = x'\hat{x}' + y'\hat{y}' = \vec{r}'$$

(a) Differentiate this relation with respect to time. Use the fact that the basis vectors $(\hat{x}, \hat{y})$ are time-independent to show that[1]

$$\vec{v} = \vec{v}' + \vec{\omega} \times \vec{r}'$$

where $\vec{v} = (dx/dt)\,\hat{x} + (dy/dt)\,\hat{y}$ is the usual velocity in the first (inertial) frame, and we have defined

$$\vec{v}' \equiv \left(\frac{dx'}{dt}\right)\hat{x}' + \left(\frac{dy'}{dt}\right)\hat{y}'$$
$$\vec{\omega} \equiv \frac{d\theta}{dt}\,\hat{z}$$

Here $\vec{v}'$ is the velocity one measures with respect to the second (rotating) frame, and $\vec{\omega}$ is the angular velocity about the z-axis;

---

[1] Use $\hat{z} \times \hat{x}' = \hat{y}'$ and $\hat{z} \times \hat{y}' = -\hat{x}'$.

(b) repeat this argument to show that with constant $\vec{\omega}$, one has

$$\frac{d\vec{v}}{dt} = \frac{d\vec{v}'}{dt} + \vec{\omega} \times \vec{v}' + \vec{\omega} \times [\vec{v}' + (\vec{\omega} \times \vec{r}')]$$

where $d\vec{v}'/dt$ is the *acceleration* one measures with respect to the second (rotating) frame

$$\frac{d\vec{v}'}{dt} \equiv \left(\frac{d^2 x'}{dt^2}\right)\hat{x}' + \left(\frac{d^2 y'}{dt^2}\right)\hat{y}'$$

(c) Hence show that when written in the second, rotating frame, Newton's second law contains additional (inertial) coriolis and centrifugal forces

$$m\frac{d\vec{v}'}{dt} = \vec{F} - 2m\vec{\omega} \times \vec{v}' - m\vec{\omega} \times (\vec{\omega} \times \vec{r}')$$

Solution to Problem 7.1

(a) We start from

$$\vec{r} = x\,\hat{x} + y\,\hat{y} = x'\hat{x}' + y'\hat{y}' = \vec{r}'$$

Differentiation of the first equality gives

$$\vec{v} = \dot{x}\,\hat{x} + \dot{y}\,\hat{y}$$

Now differentiate the last equality to obtain a second expression for $\vec{v}$

$$\frac{d\vec{r}'}{dt} = \vec{v}' + x'\frac{d\hat{x}'}{dt} + y'\frac{d\hat{y}'}{dt}$$

From the first set of relations in the problem statement above

$$\frac{d\hat{x}'}{dt} = -\dot{\theta}\,\hat{x}\sin\theta + \dot{\theta}\,\hat{y}\cos\theta$$

$$\frac{d\hat{y}'}{dt} = -\dot{\theta}\,\hat{x}\cos\theta - \dot{\theta}\,\hat{y}\sin\theta$$

Use

$$\vec{\omega} = \dot{\theta}\,\hat{z} \qquad\qquad ; \hat{z} \times \hat{x}' = \hat{y}' \qquad\qquad ; \hat{z} \times \hat{y}' = -\hat{x}'$$

It follows that the unit basis vectors satisfy the *important relations*

$$\frac{d\hat{x}'}{dt} = \dot{\theta}\,\hat{y}' = \vec{\omega} \times \hat{x}' \qquad\qquad ; \text{basis vectors}$$

$$\frac{d\hat{y}'}{dt} = -\dot{\theta}\,\hat{x}' = \vec{\omega} \times \hat{y}'$$

Therefore

$$x' \frac{d\hat{x}'}{dt} + y' \frac{d\hat{y}'}{dt} = \vec{\omega} \times \vec{r}'$$

Hence, we arrive at the important relation between the particle velocities in the two frames

$$\vec{v} = \vec{v}' + \vec{\omega} \times \vec{r}'$$

(b) This expression can now be differentiated with respect to time for constant $\vec{\omega}$ in exactly the same fashion to give the relation between the particle *accelerations* in the two frames

$$\frac{d\vec{v}}{dt} = \frac{d\vec{v}'}{dt} + \vec{\omega} \times \vec{v}' + \vec{\omega} \times [\vec{v}' + (\vec{\omega} \times \vec{r}')]$$

where $d\vec{v}'/dt$ is the *acceleration* one measures with respect to the second (rotating) frame

$$\frac{d\vec{v}'}{dt} \equiv \left(\frac{d^2 x'}{dt^2} \right) \hat{x}' + \left(\frac{d^2 y'}{dt^2} \right) \hat{y}'$$

(c) Hence when written in the second, rotating frame, Newton's second law contains additional (inertial) coriolis and centrifugal forces

$$m \frac{d\vec{v}'}{dt} = \vec{F} - 2m\vec{\omega} \times \vec{v}' - m\vec{\omega} \times (\vec{\omega} \times \vec{r}')$$

Problem 7.2 Suppose that in an inertial frame one has a particle undergoing uniform rotational motion with a radius a, an angular velocity ω, and a central force supplied by the tension τ in a string. Now view the same physical situation from a frame rotating with the same angular velocity ω. In this frame, the particle *does not move*. Use the result in Prob. 7.1 to show that the *static* situation in the rotating frame is governed by the force equation

$$-\tau \hat{r}' - m\vec{\omega} \times (\vec{\omega} \times \vec{r}') = \left(-\tau + ma\omega^2 \right) \hat{r}' = 0$$

Interpret this result.

Solution to Problem 7.2

One exerts a radial force on the particle through the tension in the string

$$\vec{F} = -\tau \hat{r}'$$

For the static situation in the rotating frame, where the particle does not move, the tension in the string must just balance the centrifugal force.

Problem 7.3 Suppose the particle in the inertial frame in Fig. 7.1 in the text is acted upon by an additional radial force $\vec{F}_{\text{ext}} = -F_{\text{ext}}\hat{r}$ and is free to also move in the radial direction. Show that the radial component of Newton's second law with the centripetal acceleration now reads

$$\mu \left(\frac{d^2r}{dt^2} - r\omega^2 \right) = -F_{\text{ext}}$$

Rewrite this as

$$\mu \frac{d^2r}{dt^2} = -F_{\text{ext}} + \mu r\omega^2$$

Now interpret the additional (inertial) *centrifugal force* on the r.h.s.[2]

Solution to Problem 7.3

If the radial distance can also change in Fig. 7.1 in the text, then there is an additional *radial acceleration*, and Eq. (7.12) is extended to read

$$\left(\frac{d\vec{v}}{dt} \right)_{\text{radial}} = \left[\frac{d^2r}{dt^2} - r \left(\frac{d\theta}{dt} \right)^2 \right] \hat{r}$$

This expression is explicitly verified in Prob. 7.7.

With a force $\vec{F}_{\text{ext}} = -F_{\text{ext}}\hat{r}$, the radial component of Newton's second law then reads

$$\mu \left[\frac{d^2r}{dt^2} - r \left(\frac{d\theta}{dt} \right)^2 \right] = -F_{\text{ext}}$$

This is rewritten as

$$\mu \frac{d^2r}{dt^2} = -F_{\text{ext}} + \mu r \left(\frac{d\theta}{dt} \right)^2$$

We now interpret the original centripetal acceleration as an additional (inertial) *centrifugal force* on the r.h.s.

Problem 7.4 A satellite is placed into a geosynchronous orbit where it sits above a given spot on the earth's equator. How high is it?[3]

---

[2]See footnote 2 on p. 44; see also Prob. 7.7.

[3]*Hint:* Recall Eq. (7.15).

Solution to Problem 7.4

Equation (7.15) relates the period for circular motion around the earth to the radius a of the orbit

$$\tau = \frac{2\pi}{\omega} = 2\pi \left(\frac{a^3}{MG}\right)^{1/2} \qquad \text{; period}$$

For stationary motion of the satellite, the period of the orbit must be that of the rotation of the earth

$$\tau_e = 1\,\text{day} \times 24\,\text{hours/day} \times 60\,\text{minutes/hour} \times 60\,\text{seconds/minute}$$
$$= 8.64 \times 10^4\,\text{sec} \qquad \text{; period of earth's rotation}$$

With the use of the constants in Prob. 6.5, the height above the center of the earth becomes

$$a = \left[\left(\frac{\tau_e}{2\pi}\right)^2 M_e G\right]^{1/3}$$

$$= \left[\left(\frac{8.64 \times 10^4\,\text{s}}{2\pi}\right)^2 (5.976 \times 10^{24}\,\text{kg})\left(6.673 \times 10^{-11}\,\frac{\text{m}^3}{\text{kg-s}^2}\right)\right]^{1/3}$$

This gives

$$a = 4.225 \times 10^7\,\text{m} = 4.225 \times 10^4\,\text{km} \qquad \text{; radius}$$

This height can be compared to the radius of the earth

$$R_e = 6.378 \times 10^6\,\text{m} = 6.378 \times 10^3\,\text{km}$$

Problem 7.5 Coulomb's law for the electrostatic force $\vec{F}_{21}$ exerted on particle 1 with charge q_1 located at position $\vec{r}_1$, by particle 2 with charge q_2 located at position $\vec{r}_2$, is given by

$$\vec{F}_{21} = \frac{q_1 q_2}{4\pi\varepsilon_0} \frac{(\vec{r}_1 - \vec{r}_2)}{|\vec{r}_1 - \vec{r}_2|^3} \qquad \text{; Coulomb's law}$$

where, in SI units, the charge is measured in coulombs (C), and

$$\frac{1}{4\pi\varepsilon_0} = 8.988 \times 10^9\,\frac{\text{Nm}^2}{\text{C}^2}$$

If q_1 and q_2 have opposite signs so that the force is *attractive*, what are the modifications of Eqs. (7.17) and (7.19) for the circular orbits?

Solution to Problem 7.5

The transition from Newton's law for gravitation to Coulomb's law for the electrostatic force is achieved through the replacement

$$-G m_1 m_2 \rightarrow \frac{q_1 q_2}{4\pi\varepsilon_0}$$

The values of the charges for an atom composed of an electron and a nucleus with Z protons are

$$q_1 = e = -1.602 \times 10^{-19}\,\text{C} \qquad ; \text{electron charge}$$
$$q_2 = -Ze \qquad\qquad\qquad\quad ; \text{nuclear charge}$$

Equation (7.17) for the energy of an atomic orbit then becomes

$$E = -\frac{Ze^2}{4\pi\varepsilon_0}\frac{1}{2a} \qquad ; \text{energy}$$

Equation (7.19) for the radius of that atomic orbit then becomes

$$a = \frac{4\pi\varepsilon_0}{Ze^2}\frac{\vec{L}^2}{\mu} \qquad ; \text{radius}$$

A useful dimensionless number to keep in mind here is the fine-structure constant

$$\alpha \equiv \frac{e^2}{(4\pi\varepsilon_0)\hbar c} = \frac{1}{137.0} \qquad ; \text{fine-structure constant}$$

Problem 7.6 (a) Consider the bead sliding without friction down the vertical hoop in Prob. 5.2. Show the normal force exerted by the hoop on the bead is

$$\vec{F}_\perp = m\left[g\cos\theta - a\dot\theta^2\right]\hat{r}$$

(b) Use energy conservation to conclude that if the bead starts from rest at the top

$$\frac{1}{2}m(a\dot\theta)^2 = mga\,(1 - \cos\theta)$$

(c) Hence, show the normal force is

$$\vec{F}_\perp = mg\,[3\cos\theta - 2]\,\hat{r}$$

(d) Conclude that the angle at which this normal force *vanishes* is given by

$$\cos \theta_0 = \frac{2}{3}$$

Solution to Problem 7.6

(a) Problem 5.2 analyzes the tangential motion of the bead on the stationary vertical hoop, where the tangential gravitational force on the bead is $mg \sin \theta$ pointing along the hoop. The gravitational force on the bead *normal* to the wire is $\vec{F}_{\text{grav}} = -mg \cos \theta \, \hat{r}$. From Prob. 7.3, the instantaneous radial centrifugal force the falling bead exerts on the hoop is $\vec{F}_{\text{centr}} = ma\dot{\theta}^2 \, \hat{r}$. By Newton's third law, the force the hoop exerts on the bead is just the *opposite* of the force the bead exerts on the hoop.[4] Thus the normal force that the hoop exerts on the bead is given by

$$\vec{F}_\perp = m \left[g \cos \theta - a\dot{\theta}^2 \right] \hat{r}$$

This normal force balances the inward gravitational force, and is reduced by the outward centrifugal force.

(b) Since there is no dissipation, the kinetic energy of the bead is just the fall in gravitational potential energy $\Delta V_{\text{grav}} = mg\Delta z$

$$\frac{1}{2} m(a\dot{\theta})^2 = mga \left(1 - \cos \theta \right)$$

(c) It follows that the normal force exerted by the hoop on the bead is

$$\vec{F}_\perp = mg \left[3 \cos \theta - 2 \right] \hat{r}$$

(d) We conclude that the angle at which this normal force *vanishes* is given by

$$\cos \theta_0 = \frac{2}{3}$$

Problem 7.7 Suppose one is working in polar coordinates (r, θ), with orthogonal unit basis vectors $(\hat{r}, \hat{\theta})$ located at (r, θ), and pointing in the direction of the increasing coordinates.

---

[4] Note that there is no radial motion of the bead.

(a) Show that if the position changes to $(r+dr, \theta+d\theta)$, the unit vectors change by

$$d\hat{r} = d\theta\,\hat{\theta} \qquad\qquad ; \; d\hat{\theta} = -d\theta\,\hat{r}$$

(b) A particle's position is given by $\vec{r} = r\,\hat{r}$. Compute $d\vec{r}$, and show the particle's *velocity* is given by

$$\vec{v} = \frac{d\vec{r}}{dt} = \dot{r}\,\hat{r} + (r\dot{\theta})\,\hat{\theta}$$

(c) Compute $d\vec{v}$, and show the particle's *acceleration* is given by

$$\vec{a} = \frac{d\vec{v}}{dt} = (\ddot{r} - r\dot{\theta}^2)\,\hat{r} + \frac{1}{r}\frac{d}{dt}(r^2\dot{\theta})\,\hat{\theta}$$

Solution to Problem 7.7

(a) In Prob. 7.1, we studied the transformation to a rotating coordinate system. A crucial step in part (a) of the solution to that problem was the relation between the unit basis vectors

$$\frac{d\hat{x}'}{dt} = \dot{\theta}\,\hat{y}' \qquad\qquad ; \; \frac{d\hat{y}'}{dt} = -\dot{\theta}\,\hat{x}'$$

Now associate these unit vectors with those in polar coordinates

$$\hat{x}' = \hat{r} \qquad\qquad ; \; \hat{y}' = \hat{\theta}$$

Multiplication of the previous relations by dt shows that under an infinitesimal rotation $\theta \to \theta + d\theta$, these unit vectors change by

$$d\hat{r} = d\theta\,\hat{\theta} \qquad\qquad ; \; d\hat{\theta} = -d\theta\,\hat{r}$$

There is no change in these unit vectors under a simple increase in the radial coordinate $r \to r + dr$.

(b) A particle's position is given by $\vec{r} = r\,\hat{r}$. Let us compute $d\vec{r}$

$$d\vec{r} = dr\,\hat{r} + r\,d\hat{r} = dr\,\hat{r} + rd\theta\,\hat{\theta}$$

The particle's velocity in polar coordinates then follows as

$$\vec{v} = \frac{d\vec{r}}{dt} = \dot{r}\,\hat{r} + (r\dot{\theta})\,\hat{\theta}$$

(c) Now compute $d\vec{v}$

$$d\vec{v} = (d\dot{r})\hat{r} + \dot{r}\,d\hat{r} + (dr\,\dot{\theta} + r\,d\dot{\theta})\,\hat{\theta} + (r\dot{\theta})\,d\hat{\theta}$$
$$= \left(d\dot{r} - r\dot{\theta}\,d\theta\right)\hat{r} + \left(dr\,\dot{\theta} + r\,d\dot{\theta} + \dot{r}\,d\theta\right)\hat{\theta}$$

Hence the particle's acceleration in polar coordinates is

$$\vec{a} = \frac{d\vec{v}}{dt} = (\ddot{r} - r\dot{\theta}^2)\,\hat{r} + (2\dot{r}\dot{\theta} + r\ddot{\theta})\,\hat{\theta}$$

which is rewritten as

$$\vec{a} = \frac{d\vec{v}}{dt} = (\ddot{r} - r\dot{\theta}^2)\,\hat{r} + \frac{1}{r}\left[\frac{d}{dt}(r^2\dot{\theta})\right]\hat{\theta}$$

Problem 7.8 Suppose one is in a closed laboratory in space, away from all external fields. Recall the results in chapter 3 and Prob. 7.1.

(a) Make an argument that there is no mechanics experiment one can do that will determine whether or not the laboratory is moving with constant velocity relative to the fixed stars;

(b) Show there *is* an experiment one can do that will determine whether or not the laboratory is *rotating* with respect to the fixed stars.

Solution to Problem 7.8

(a) If we cannot see out, and all we can do is a mechanics experiment inside the closed laboratory, then there is *no way* we can tell whether or not we are moving with a uniform velocity relative to the fixed stars. All such frames are inertial, in which Newton's laws hold, and there are *no additional inertial forces*.

(b) It is thought-provoking that there *is* a way we can tell from experiments done inside such a laboratory whether or not we are *rotating* with respect to the fixed stars, since the inertial coriolis and centrifugal forces of Prob. 7.1 *are present* in the rotating laboratory.

Chapter 8

System of Particles

Problem 8.1 In Eq. (8.28) we identified the moment of inertia for a cylinder rotating about its figure axis as

$$I_0 = \rho \int \vec{r}^{\,2}\, dv \qquad ; \text{ moment of inertia of cylinder}$$

Here $\vec{r}$ is a vector transverse to the symmetry axis running out to the volume element dv. The moment of inertia I_0 about the symmetry axis in the C-M frame is entirely a property of the body. Suppose that instead of about the symmetry axis, we want to compute the moment of inertia about a *parallel* axis running along the l.h.s. of the cylinder in Fig. 8.1 below. Let $\vec{u}$ be a vector transverse to that axis and again running out to the volume element dv. Write

$$\vec{u} = \vec{l} + \vec{r}$$

where $\vec{l}$ is a vector running in to the symmetry axis, and transverse to it. Now evaluate the moment of inertia about this new axis of rotation

$$I = \rho \int \vec{u}^{\,2}\, dv = \rho \int \left(\vec{r} + \vec{l} \right)^2 dv$$

This is rewritten as

$$I = \rho \int \left(\vec{r}^{\,2} + 2\vec{r} \cdot \vec{l} + \vec{l}^{\,2} \right) dv$$

Show the integration over the term linear in $\vec{r}$ *vanishes by symmetry*, and hence

$$I = I_0 + Ml^2 \qquad ; \text{ parallel-axis theorem}$$

This is a simple example of what is known as the *parallel-axis theorem*. The new moment of inertial about the new axis is the old moment of inertia about the symmetry axis plus Ml^2, where l is the transverse distance between the axes.

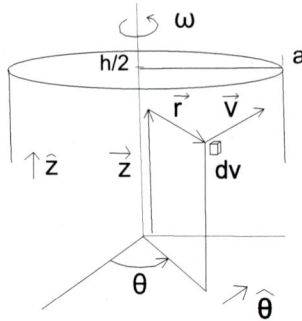

Fig. 8.1   Solid right circular cylinder of height h and radius a rotating about the symmetry axis (z-axis) with angular velocity ω. The volume element dv lies a perpendicular distance $\vec{r}$ from the axis of rotation. The mass density of the cylinder is ρ.

Solution to Problem 8.1

Consider a cylindrical slab of thickness dz in Fig. 8.1, and a parallel axis running down the l.h.s. of the cylinder. The vector $\vec{u}$ then runs from that axis to the volume dv, and $\vec{l}$ from that axis to the symmetry axis, so that on the slab

$$\vec{u} = \vec{l} + \vec{r}$$

We also use cylindrical coordinates so that

$$dv = r\, dr\, d\theta\, dz$$

The moment of inertia of the cylinder about the new parallel axis is then

$$I = \rho \int \vec{u}^{\,2}\, dv = \rho \int \left(\vec{r} + \vec{l}\,\right)^2 dv$$
$$= \rho \int \left(\vec{r}^{\,2} + 2\vec{r}\cdot\vec{l} + \vec{l}^{\,2}\right) dv$$

Now the vector $\vec{l}$ is fixed, and it is easy to argue from symmetry, or to show by explicit integration,[1] that

$$\int \left(\vec{r} \cdot \vec{l} \right) dv = 0$$

It follows that

$$I = I_0 + Ml^2 \qquad ; \text{ parallel-axis theorem}$$

This is a simple example of what is known as the *parallel-axis theorem*.

Problem 8.2 Show the kinetic energy of the rotating cylinder in Fig. 8.1 above is

$$T' = \frac{1}{2} I \omega^2 \qquad ; \text{ rotating cylinder}$$

Solution to Problem 8.2

The velocity of each little element in Fig. 8.1 above is $\vec{v} = d\vec{r}/dt = (r\omega)\,\hat{\theta}$. Thus, in line with Eq. (8.28) in the text, the total kinetic energy of the rotating cylinder is

$$T' = \rho \int dv \, \frac{1}{2} \left(r\omega \, \hat{\theta} \right)^2 = \frac{1}{2} \omega^2 \rho \int r^2 \, dv$$

$$= \frac{1}{2} I \omega^2$$

Problem 8.3 Show that the moment of inertia for a hoop with radius a for rotation about a normal axis through its center is

$$I = Ma^2 \qquad ; \text{ hoop}$$

Solution to Problem 8.3

This one is easy. All the mass is uniformly distributed around a circle a distance a away from the axis of rotation, which is normal to it. Hence

$$I = \int \rho r^2 \, dv = Ma^2$$

Problem 8.4 Show that the moment of inertia for a sphere with radius a for rotation about a diameter is

$$I = \frac{2}{5} Ma^2 \qquad ; \text{ sphere}$$

---

[1] Use $\int_0^{2\pi} \sin\theta \, d\theta = \int_0^{2\pi} \cos\theta \, d\theta = 0$.

Solution to Problem 8.4

Introduce spherical coordinates in the sphere with volume element

$$dv = r^2\, dr \sin\theta\, d\theta\, d\phi$$

The mass of the sphere is then

$$M = \rho \int_0^a r^2\, dr \int_0^\pi \sin\theta\, d\theta \int_0^{2\pi} d\phi$$

$$= \rho \frac{2\pi a^3}{3} \int_{-1}^1 dx \qquad ; x \equiv \cos\theta$$

$$= \rho \frac{4\pi a^3}{3}$$

where we have introduced $x = \cos\theta$ in the second line.

The moment of inertia of the sphere about the z-axis then has an additional $(r\sin\theta)^2$ in the integrand, the square of the perpendicular distance of the volume element dv from the z-axis,

$$I = \rho \int_0^a r^2\, dr \int_0^\pi \sin\theta\, d\theta\, (r\sin\theta)^2 \int_0^{2\pi} d\phi$$

$$= \rho \frac{2\pi a^5}{5} \int_{-1}^1 (1-x^2)\,dx \qquad ; x \equiv \cos\theta$$

$$= \rho \frac{2\pi a^5}{5} \frac{4}{3}$$

$$= \rho \frac{4\pi a^3}{3} \frac{2a^2}{5}$$

It follows that

$$I = \frac{2}{5} M a^2 \qquad ; \text{sphere}$$

Problem 8.5 Show that the moment of inertia for a hoop with radius a for rotation about a normal axis through its *rim* is[2]

$$I = 2Ma^2 \qquad ; \text{hoop rim}$$

Solution to Problem 8.5

This one is also easy once we have the moment of inertia of the hoop in Prob. 8.3 and the parallel axis theorem in Prob. 8.1. The new axis is now

---

[2] *Hint*: Recall Prob. 8.1.

parallel to the one that is through the center of the hoop and perpendicular to it. It is a distance a away. Thus

$$I = Ma^2 + Ma^2 = 2Ma^2$$

Problem 8.6 A spool of mass M, radius a, and moment of inertia $I = Ma^2/2$ stands on end and slides with no friction on a horizontal table here in the lab. It is wound with a string which connects to a mass m hanging over the edge of the table. Let x be the linear displacement of the spool, θ the angle of rotation of the spool, and z the distance down of the hanging mass.

(a) Show the equation of motion of the C-M of the spool is

$$M\ddot{x} = mg$$

(b) Show the unwinding of the string is governed by the equation

$$I\ddot{\theta} = amg$$

(c) Hence show the increase in the length z is governed by the equation

$$\ddot{z} = \ddot{x} + a\ddot{\theta} = mg \left(\frac{1}{M} + \frac{a^2}{I} \right)$$
$$= \frac{3mg}{M}$$

(d) Integrate this last relation to find $z(t)$.

Solution to Problem 8.6

This is a simple problem, but it nicely illustrates rigid-body motion.

(a) The center-of-mass (C-M) moves as if all the external forces acting on the body were applied to the C-M. If x locates that position, then the hanging mass exerts a gravitational force mg on the spool transmitted through the string tension, and[3]

$$M\ddot{x} = mg$$

(b) The string tension acts at the side of the spool, a distance a away from the symmetry axis, and it exerts a torque about the symmetry axis of mga. Hence the rate of change of angular momentum of the spool about

---

[3]It is assumed here that $m \ll M$, and the tension in the string is $\tau \approx mg$.

the symmetry axis is

$$I\ddot{\theta} = amg$$

(c) The length of string that unwinds from the spool is $a\theta$. Hence, the increase in the length z of the hanging mass is governed by the equation

$$\ddot{z} = \ddot{x} + a\ddot{\theta} = mg\left(\frac{1}{M} + \frac{a^2}{I}\right)$$

$$= \frac{3mg}{M}$$

(d) If the spool starts from rest at the origin, the solution to this equation is

$$z = \frac{1}{2}\left(\frac{3mg}{M}\right)t^2$$

Problem 8.7 This problem concerns the *statics* of the rigid-body discussed in the text. Suppose the rigid body is *at rest* in an inertial frame, and consider any other orthogonal coordinate system in that inertial frame.

(a) Show that in the second frame

$$\sum_i \vec{F}_i^{(e)} = 0 \qquad ; \text{statics}$$

Note that this holds no matter where the individual external forces $\vec{F}_i^{(e)}$ act on the body;

(b) Let $\vec{\rho}_i$ denote the location of the ith constituent of the rigid body in the second frame. Show that the net *torque* in that frame must also vanish

$$\sum_i \vec{\rho}_i \times \vec{F}_i^{(e)} = 0 \qquad ; \text{statics}$$

Solution to Problem 8.7

(a) It is the sum of all the external forces acting on a rigid body that governs the motion of the center-of-mass (C-M) in Eq. (8.7). If the C-M is at *at rest*, then this net force must vanish

$$\sum_i \vec{F}_i^{(e)} = 0 \qquad ; \text{statics}$$

(b) It is the net torque from these external forces that determines the time rate of change of the angular momentum of a rigid body in the C-M frame in Eq. (8.26). If this angular momentum is constant, or absent, then the net torque must also vanish

$$\sum_i \vec{\rho}_i \times \vec{F}_i^{(e)} = 0 \qquad ; \text{ statics}$$

Problem 8.8 (a) It is desired to support a heavy mass M in the earth's gravitational field. A round "pulley" is attached to the ceiling and a rope is extended down to another pulley attached to the weight, and then back to the ceiling through another pulley, and then back down to the weight, and so on, until there are N segments of the rope supporting the weight. The rope then extends out to an operator who applies a force to the rope which creates a uniform tension in it. What force must the operator apply to the rope to just keep the weight lifted (neglect the weight of the support system)? Why? [4]

(b) Two people sit at opposite ends of a teeter-totter. The first has a mass of M and sits a distance d from the fulcrum. The second has a mass of $M/2$. Where must the second person sit to just balance the teeter-totter? Why?

Solution to Problem 8.8

(a) Suppose that with the block and tackle, the weight is supported by N segments of the rope, each of which carries an equal part of the support. Then the uniform tension τ in the rope is

$$N\tau = Mg$$
$$\text{or;} \qquad \tau = \frac{1}{N} Mg$$

If I hold the end of the rope in my hand, the tension I must support is reduced by a factor of N from the weight of the mass. Alternatively, the distance through which I must pull the rope to do enough work to increase its potential energy is *increased* by a factor of N.

(b) In the static situation, the torques about the axis through the fulcrum of the teeter-totter, and perpendicular to it, must balance. Therefore

$$Mgd = \frac{Mg}{2} 2d$$

Hence the second person must sit a distance $2d$ from the fulcrum.

---

[4]The pulleys may be assembled into "blocks" of this *block and tackle* system. Note the rope must be pulled a distance Nd by the operator to raise the weight a distance d.

Chapter 9

Generalized Coordinates

Problem 9.1 (a) Sketch the motion of the uncoupled normal modes in Eqs. (9.43) by setting one or the other of the coordinates (X, x) equal to zero;

(b) Can you understand the restoring force in each case?

(c) Write the general solution for the normal modes in Eqs. (9.43).

Solution to Problem 9.1

(a) We shall simply refer to Fig. 9.4 in the text. One normal mode leaves the inter-particle distance unchanged, while the C-M oscillates back and forth, with the restoring force coming from the two end springs.

The second normal mode leaves the position of the C-M unchanged, while the two particles oscillate against each other. In this case, the restoring force is increased by the spring between the masses.

(b) In this first case, the restoring force on $m\ddot{X}$ is just the spring constant κ from the end springs [see Eqs. (9.43)], and the normal-mode frequency is $\Omega_1^2 = \kappa/m$.

In the second case, the restoring force on $m\ddot{x}$ is the spring constant κ from the end springs, and 2κ from the connecting spring, hence the normal-mode frequency is $\Omega_2^2 = 3\kappa/m$.

(c) The general solution for the motion of each mass is just a linear combination of normal modes [see Eqs. (9.42)]

$$x_1 = X - \frac{1}{2}x = A_1 \cos(\Omega_1 t + \phi_1) - \frac{1}{2}A_2 \cos(\Omega_2 t + \phi_2)$$

$$x_2 = X + \frac{1}{2}x = A_1 \cos(\Omega_1 t + \phi_1) + \frac{1}{2}A_2 \cos(\Omega_2 t + \phi_2)$$

Problem 9.2 A pendulum of length l and mass m hangs down from a fixed support. A second pendulum of the same length and mass hangs down from the first. Take the angles (θ_1, θ_2) of displacement from the vertical as generalized coordinates.

(a) Show the lagrangian for the coupled system is

$$L(\dot{\theta}_1, \dot{\theta}_2, \theta_1, \theta_2) = \frac{ml^2}{2}[2\dot{\theta}_1^2 + \dot{\theta}_2^2 + 2\dot{\theta}_1\dot{\theta}_2 \cos(\theta_2 - \theta_1)] +$$
$$mgl(2\cos\theta_1 + \cos\theta_2)$$

(b) Show that the linearized form of Lagrange's equations read

$$2\ddot{\theta}_1 + \ddot{\theta}_2 = -2\omega_0^2\theta_1 \qquad ; \ \omega_0^2 \equiv \frac{g}{l}$$
$$\ddot{\theta}_2 + \ddot{\theta}_1 = -\omega_0^2\theta_2$$

(c) Look for normal modes where everything oscillates with the same angular velocity so that

$$\ddot{\theta}_2 = -\omega^2\theta_2 \qquad ; \ \ddot{\theta}_1 = -\omega^2\theta_1$$

Show the normal-mode frequencies are given by

$$\omega^2 = \omega_0^2\left(2 \pm \sqrt{2}\right)$$

Solution to Problem 9.2

(a) Suppose the pendulums oscillate in the (x, z)-plane, with z pointing down. The cartesian coordinates of the pendulums are then

$$x_1 = l\sin\theta_1 \qquad ; \ x_2 = l\sin\theta_1 + l\sin\theta_2$$
$$z_1 = l\cos\theta_1 \qquad ; \ z_2 = l\cos\theta_1 + l\cos\theta_2$$

Differentiate these relations

$$\dot{x}_1 = l\dot{\theta}_1 \cos\theta_1 \qquad ; \ \dot{x}_2 = l\dot{\theta}_1 \cos\theta_1 + l\dot{\theta}_2 \cos\theta_2$$
$$\dot{z}_1 = -l\dot{\theta}_1 \sin\theta_1 \qquad ; \ \dot{z}_2 = -l\dot{\theta}_1 \sin\theta_1 - l\dot{\theta}_2 \sin\theta_2$$

Compute

$$\dot{x}_1^2 + \dot{z}_1^2 + \dot{x}_2^2 + \dot{z}_2^2 = (l\dot{\theta}_1)^2 + (l\dot{\theta}_1)^2 + (l\dot{\theta}_2)^2 +$$
$$2l^2(\dot{\theta}_1\dot{\theta}_2)[\cos\theta_1 \cos\theta_2 + \sin\theta_1 \sin\theta_2]$$

Hence the kinetic energy of the coupled pendulums is

$$T = \frac{m}{2}\left[(l\dot{\theta}_1)^2 + (l\dot{\theta}_1)^2 + (l\dot{\theta}_2)^2 + 2l^2(\dot{\theta}_1\dot{\theta}_2)\cos(\theta_2 - \theta_1)\right]$$

The gravitational potential energy is

$$V = -mg\left[2l\cos\theta_1 + l\cos\theta_2\right]$$

The lagrangian for the coupled systems is therefore

$$L(\dot{\theta}_1, \dot{\theta}_2, \theta_1, \theta_2) = \frac{ml^2}{2}[2\dot{\theta}_1^2 + \dot{\theta}_2^2 + 2\dot{\theta}_1\dot{\theta}_2\cos(\theta_2 - \theta_1)] +$$
$$mgl(2\cos\theta_1 + \cos\theta_2)$$

(b) The *linearized* form of the lagrangian is[1]

$$L(\dot{\theta}_1, \dot{\theta}_2, \theta_1, \theta_2) \doteq \frac{ml^2}{2}[2\dot{\theta}_1^2 + \dot{\theta}_2^2 + 2\dot{\theta}_1\dot{\theta}_2] + \frac{mgl}{2}[6 - 2\theta_1^2 - \theta_2^2]$$

The linearized Lagrange equations are then

$$2\ddot{\theta}_1 + \ddot{\theta}_2 = -2\omega_0^2\theta_1 \qquad ; \; \omega_0^2 \equiv \frac{g}{l}$$
$$\ddot{\theta}_2 + \ddot{\theta}_1 = -\omega_0^2\theta_2$$

(c) If we look for the normal modes of part (c), the above reduce to coupled, linear, homogeneous equations

$$\left(2\omega_0^2 - 2\omega^2\right)\theta_1 - \omega^2\theta_2 = 0$$
$$-\omega^2\theta_1 + \left(\omega_0^2 - \omega^2\right)\theta_2 = 0$$

These equations will only have a non-trivial solution if the determinant of the coefficients vanishes

$$\omega^4 - 4\omega_0^2\omega^2 + 2\omega_0^4 = 0$$

The solutions to this quadratic equation for ω^2 yield the normal-mode frequencies of the coupled systems as

$$\omega^2 = \omega_0^2(2 \pm \sqrt{2})$$

Problem 9.3 A particle of mass m is constrained to move without friction on the surface of a right cylinder of radius a oriented so the symmetry axis of the cylinder is perpendicular to the earth's surface. Let z be the height along the symmetry axis and θ be the polar angle with cylindrical coordinates (see Fig. 8.2 in the text).

---

[1]That is, it leads to *linearized* Lagrange equations.

(a) Show the lagrangian, including the gravitational potential, is

$$L(\dot{z}, \dot{\theta}, z) = \frac{m}{2}\left[\dot{z}^2 + (a\dot{\theta})^2\right] - mgz$$

(b) Choose (z, θ) as generalized coordinates. Show Lagrange's equations are

$$m\ddot{z} = -mg$$

$$\frac{d}{dt}(ma^2\dot{\theta}) = 0$$

(c) Suppose the particle is projected from the origin $(z, \theta) = (0, 0)$ at time $t = 0$ with an initial velocity

$$v_{0z} = v_0 \cos\phi \qquad ; \; v_{0\theta} = v_0 \sin\phi$$

Show the solution to the equations of motions is

$$z = (v_0 \cos\phi)t - \frac{g}{2}t^2$$

$$a\theta = (v_0 \sin\phi)t$$

(d) Assume the angle ϕ is finite. Show the orbit $z(\theta)$ is given by

$$z(\theta) = (a\theta)\cot\phi - \frac{g}{2}\left(\frac{a\theta}{v_0 \sin\phi}\right)^2$$

Solution to Problem 9.3

(a) It is always best to start in cartesian coordinates. The particle coordinates are then

$$x = a\cos\theta \qquad ; \; y = a\sin\theta \qquad ; \; z = z$$

It then follows as in the text that the kinetic energy of the particle is

$$T = \frac{m}{2}\left[\dot{x}^2 + \dot{y}^2 + \dot{z}^2\right] = \frac{m}{2}\left[(a\dot{\theta})^2 + \dot{z}^2\right]$$

The gravitational potential energy follows from the height in the z-direction

$$V = mgz$$

The lagrangian is therefore given by

$$L = T - V = \frac{m}{2}\left[\dot{z}^2 + (a\dot{\theta})^2\right] - mgz$$

(b) Choose (z, θ) as generalized coordinates so that we have $L(\dot{z}, \dot{\theta}, z)$. Lagrange's equations are then

$$\frac{d}{dt}\left(\frac{\partial L}{\partial \dot{z}}\right) - \frac{\partial L}{\partial z} = m\ddot{z} + mg = 0$$

$$\frac{d}{dt}\left(\frac{\partial L}{\partial \dot{\theta}}\right) - \frac{\partial L}{\partial \theta} = \frac{d}{dt}(ma^2\dot{\theta}) = 0$$

(c) Suppose the particle is projected from the origin $(z, \theta) = (0, 0)$ at time $t = 0$ with an initial velocity

$$v_{0z} = v_0 \cos\phi \qquad ; \quad v_{0\theta} = v_0 \sin\phi$$

It follows exactly as in the text that the solution to the equations of motions is then

$$z = (v_0 \cos\phi)t - \frac{g}{2}t^2$$
$$a\theta = (v_0 \sin\phi)t$$

(d) Assume the angle ϕ is finite. Solve the second equation for t, and then substitute it in the first relation to obtain the orbit $z(\theta)$ on the curved surface

$$z(\theta) = (a\theta)\cot\phi - \frac{g}{2}\left(\frac{a\theta}{v_0 \sin\phi}\right)^2$$

Chapter 10

Hamilton's Principle

Problem 10.1 If there are no singularities lurking around, a well-behaved function $f(x)$ has a power-series expansion about a neighboring point x_0 of the form

$$f(x) = \sum_{n=0}^{\infty} a_n (x - x_0)^n \qquad ; \text{ Taylor series}$$

This is known as a *Taylor series*. Show that the first two interesting coefficients in this series are obtained from the derivatives of the function at x_0 according to[1]

$$a_1 = \left[\frac{df(x)}{dx} \right]_{x=x_0}$$

$$a_2 = \frac{1}{2!} \left[\frac{d^2 f(x)}{dx^2} \right]_{x=x_0}$$

Solution to Problem 10.1

Assume the function $f(x)$ is well-behaved in the region surrounding the point x_0 and that we can make a power series expansion in the distance from that point

$$f(x) = \sum_{n=0}^{\infty} a_n (x - x_0)^n$$

$$= a_0 + (x - x_0)a_1 + (x - x_0)^2 a_2 + (x - x_0)^3 a_3 + \cdots$$

Assume also that this series can be differentiated term by term. It is then

---

[1] Note that $a_0 = f(x_0)$. The *Taylor series* also plays a central role in our analysis.

evident that

$$f(x_0) = a_0$$

$$\left[\frac{df(x)}{dx}\right]_{x=x_0} = a_1$$

$$\left[\frac{d^2 f(x)}{dx^2}\right]_{x=x_0} = 2! \, a_2 \qquad ; \text{ etc.}$$

Hence, in general, one arrives at

$$f(x) = \sum_{n=0}^{\infty} a_n (x - x_0)^n \qquad ; \text{ Taylor series}$$

$$a_n = \frac{1}{n!} \left[\frac{d^n f(x)}{dx^n}\right]_{x=x_0}$$

This is a *Taylor series*. It expresses the function $f(x)$ in terms of its value and all its derivatives at the neighboring point x_0.

Problem 10.2 This problem presents an application of the calculus of variations.

(a) Show that the time it takes for a light ray to propagate a short distance in the (x, y)-plane is [recall Eq. (10.2)]

$$dt = \frac{1}{c}\sqrt{1 + y'^2}\, dx$$

where c is the velocity of light in the medium in which it is propagating;

(b) Suppose a light ray starts at a distance h_1 above the origin in a first medium, propagates to a point x on the x-axis, and then enters a second medium when after a total distance l on the x-axis, it ends up a distance h_2 below it. *Fermat's principle* states that the path the light takes is that which minimizes the transit time. Use the arguments in the text to show that the path will be a straight line down to x, and then a second straight-line down to $-h_2$, and the total time taken, τ, is

$$\tau = \frac{1}{c_1}\sqrt{h_1^2 + x^2} + \frac{1}{c_2}\sqrt{h_2^2 + (l - x)^2}$$

(c) Find the position x that minimizes the total transit time, and show

$$\frac{1}{c_1}\frac{x}{\sqrt{h_1^2 + x^2}} = \frac{1}{c_2}\frac{(l - x)}{\sqrt{h_2^2 + (l - x)^2}} \qquad ; \text{ Fermat's principle}$$

(d) Introduce the index of refraction $n_i = c/c_i$ in each medium, where c is the velocity of light in vacuum, and the angle θ_i with respect to the normal to the interface in each medium. Show the result in part (c) can be written as

$$n_1 \sin \theta_1 = n_2 \sin \theta_2 \qquad ;\text{ Snell's law}$$

This is *Snell's law* of refraction.

Solution to Problem 10.2

(a) Equation (10.2) states that the element of infinitesimal distance between two points in the (x, y)-plane is

$$ds = \sqrt{1 + y'^2}\, dx$$

If c is the velocity of light in the medium in which it is propagating, then the time it takes to cover that distance is $c\, dt = ds$. Thus

$$dt = \frac{1}{c}\sqrt{1 + y'^2}\, dx$$

(b) We used the calculus of variations to show that the shortest distance between two points is a straight line. Hence, the shortest time for light to transverse the configuration described in part (b) above is

$$\tau = \frac{1}{c_1}\sqrt{h_1^2 + x^2} + \frac{1}{c_2}\sqrt{h_2^2 + (l - x)^2}$$

(c) Now minimize this expression with respect to the choice of x to satisfy Fermat's principle. To find the minimum, set the derivative with respect to x equal to zero

$$\frac{1}{c_1}\frac{x}{\sqrt{h_1^2 + x^2}} = \frac{1}{c_2}\frac{(l - x)}{\sqrt{h_2^2 + (l - x)^2}} \qquad ;\text{ Fermat's principle}$$

(d) Introduce the index of refraction $n_i = c/c_i$ in each medium, where c is the velocity of light in vacuum, and the angle θ_i with respect to the normal to the interface in each medium. It follows that the result in part (c) can be written as

$$n_1 \sin \theta_1 = n_2 \sin \theta_2 \qquad ;\text{ Snell's law}$$

This is *Snell's law* of refraction.

Problem 10.3 Recall Prob. 1.4. Use a Taylor series to show that for small x

$$(1+x)^n \approx 1 + nx + \frac{n(n-1)}{2!}x^2 + \cdots$$

for both integral and half-integral n.

Solution to Problem 10.3

Problem 1.4 establishes the fact that the relation

$$\frac{d}{dt}(t^n) = nt^{n-1}$$

holds for all integer and half-integer n. Now make a Taylor series expansion of the function

$$f(x) = (1+x)^n$$

about the point $x_0 = 0$. With the chain rule, one has

$$f(0) = 1$$

$$\left[\frac{df(x)}{dx}\right]_{x=0} = n$$

$$\frac{1}{2!}\left[\frac{d^2 f(x)}{dx^2}\right]_{x=0} = \frac{1}{2!}n(n-1)$$

Hence the Taylor series gives us for both integer and half-integer n

$$(1+x)^n = 1 + nx + \frac{n(n-1)}{2!}x^2 + \cdots$$

This is an extremely useful expansion for small x.

Chapter 11

Lagrangian Dynamics

Problem 11.1 (a) Show that the lagrangian for the simple harmonic oscillator is

$$L = \frac{m}{2}\left(\frac{dx}{dt}\right)^2 - \frac{\kappa}{2}x^2 \qquad \text{; simple harmonic oscillator}$$

Write Lagrange's equation and show it reproduces Newton's law;

(b) Show the canonical momentum is

$$p = m\frac{dx}{dt}$$

Interpret this result.

(c) Construct the hamiltonian and show

$$H(p,q) = \frac{p^2}{2m} + \frac{1}{2}\kappa q^2$$

Solution to Problem 11.1

(a) The kinetic and potential energies for the one-dimensional oscillator are

$$T(\dot{x}) = \frac{1}{2}m\left(\frac{dx}{dt}\right)^2 \qquad ; V(x) = \frac{1}{2}\kappa x^2$$

Thus the lagrangian is

$$L(\dot{x},x) = T(\dot{x}) - V(x) = \frac{1}{2}m\left(\frac{dx}{dt}\right)^2 - \frac{1}{2}\kappa x^2$$

Lagrange's equation then gives

$$\frac{d}{dt}\left(\frac{\partial L}{\partial \dot{x}}\right) - \frac{\partial L}{\partial x} = m\frac{d^2x}{dt^2} + \kappa x = 0$$

This is Newton's second law for the oscillator.

(b) The canonical momentum is defined by

$$p \equiv \frac{\partial L}{\partial \dot{x}} = m\frac{dx}{dt}$$

This is actual momentum of the particle.

(c) The hamiltonian is defined by

$$H \equiv p\frac{dx}{dt} - L = \frac{1}{2}m\left(\frac{dx}{dt}\right)^2 + \frac{1}{2}\kappa x^2 = E$$

This is evidently the energy. When written in terms of the canonical momentum, with the coordinate x defined as q, one has

$$H(p,q) = \frac{p^2}{2m} + \frac{1}{2}\kappa q^2 \qquad ; x \equiv q$$

Problem 11.2 A particle of mass m slides without friction inside the bottom half of a spherical bowl of radius a whose symmetry axis is vertical at the earth's surface. Let the z-axis point down, and introduce spherical coordinates

$$x = a\sin\theta\cos\phi \qquad ; y = a\sin\theta\sin\phi \qquad ; z = a\cos\theta$$

(a) Show the kinetic and potential energies are given by

$$T = \frac{m}{2}\left(\dot{x}^2 + \dot{y}^2 + \dot{z}^2\right) = \frac{ma^2}{2}\left(\dot{\theta}^2 + \dot{\phi}^2\sin^2\theta\right)$$
$$V = -mga\cos\theta$$

Hence, show that with generalized coordinates (θ, ϕ) the lagrangian is

$$L(\dot{\theta}, \dot{\phi}, \theta, \phi) = \frac{ma^2}{2}\left(\dot{\theta}^2 + \dot{\phi}^2\sin^2\theta\right) + mga\cos\theta$$

(b) Show that Lagrange's equations are

$$\frac{d}{dt}\left(ma^2\dot{\phi}\sin^2\theta\right) = 0 \qquad\qquad ; \phi \text{ eqn}$$
$$ma^2\ddot{\theta} = ma^2\dot{\phi}^2\sin\theta\cos\theta - mga\sin\theta \qquad ; \theta \text{ eqn}$$

(c) Show that there is an orbit where ϕ is constant and θ obeys the pendulum equation

$$\ddot{\theta} = -\frac{g}{a} \sin \theta \qquad ; \ \phi = \text{constant}$$

(d) Show that there is another orbit where θ is constant and $\dot{\phi}$ is given by

$$\dot{\phi}^2 = \frac{g}{a \cos \theta} \qquad ; \ \theta = \text{constant}$$

Interpret this last result in terms of the constraint force exerted by the wall of the bowl on the mass.[1]

Solution to Problem 11.2

(a) Take the time derivative of the transformation equations

$$\dot{x} = a[\dot{\theta} \, \cos\theta \cos\phi - \dot{\phi} \, \sin\theta \sin\phi]$$
$$\dot{y} = a[\dot{\theta} \, \cos\theta \sin\phi + \dot{\phi} \, \sin\theta \cos\phi]$$
$$\dot{z} = -a\dot{\theta} \, \sin\theta$$

Hence, the kinetic energy is[2]

$$T = \frac{m}{2} \left(\dot{x}^2 + \dot{y}^2 + \dot{z}^2 \right) = \frac{ma^2}{2} \left(\dot{\theta}^2 + \dot{\phi}^2 \sin^2\theta \right)$$

With z pointing down, the potential energy is

$$V = -mgz = -mga \cos\theta$$

Thus, with the generalized coordinates (θ, ϕ), the lagrangian is

$$L(\dot{\theta}, \dot{\phi}, \theta, \phi) = \frac{ma^2}{2} \left(\dot{\theta}^2 + \dot{\phi}^2 \sin^2\theta \right) + mga \cos\theta$$

(b) It follows that Lagrange's equations are

$$\frac{d}{dt} \left(ma^2 \dot{\phi} \sin^2\theta \right) = 0 \qquad\qquad ; \ \phi \text{ eqn}$$
$$ma^2\ddot{\theta} = ma^2 \dot{\phi}^2 \sin\theta \cos\theta - mga \sin\theta \qquad ; \ \theta \text{ eqn}$$

---

[1] *Hint*: It must be normal to the surface.

[2] A quick way to obtain this result is to realize that the velocity is expressed as $\vec{v} = (a\dot{\theta})\hat{\theta} + (a \sin\theta \, \dot{\phi})\hat{\phi}$ in the tangent plane.

(c) If ϕ is constant, corresponding to motion in a transverse plane passing through the z-axis, then θ obeys the pendulum equation

$$\ddot{\theta} = -\frac{g}{a}\sin\theta \qquad ; \phi = \text{constant}$$

(d) If θ is constant, corresponding to horizontal circular motion in the bowl around the z-axis, then $\dot{\phi}$ is given by

$$\dot{\phi}^2 = \frac{g}{a\cos\theta} \qquad ; \theta = \text{constant}$$

In this last case, the frictionless constraint force exerted by the wall of the bowl on the mass cannot exert any tangential force, and the upward tangential centrifugal force is just balanced by the downward tangential gravitational force

$$(F_{\text{grav}})_\| = mg\sin\theta$$
$$(F_{\text{cent}})_\| = m(a\sin\theta\,\dot{\phi}^2)\cos\theta$$

Problem 11.3 (a) Introduce the constant angular momentum in the previous problem

$$L_z = ma^2\dot{\phi}\sin^2\theta \qquad ; \text{angular momentum}$$

Rewrite the equation for θ to show that

$$\ddot{\theta} = \frac{L_z^2}{m^2 a^4}\frac{\cos\theta}{\sin^3\theta} - \frac{g}{a}\sin\theta$$

(b) Look for an equilibrium solution to this equation with $\theta = \theta_0 = $ constant. Show this reproduces the result in Eq. (17.138);

(c) Now look for small oscillations about this equilibrium value at fixed angular momentum L_z

$$\theta = \theta_0 + \eta \qquad ; |\eta| \ll 1$$

Discuss the stability of these circular orbits.

Solution to Problem 11.3

(a) The first of Lagrange's equations in the previous problem states that the angular momentum along the z-axis is a constant of the motion[3]

$$L_z = m(a\sin\theta)^2\dot{\phi} = \text{constant}$$

---

[3] There are no torques about the z-axis.

Use this relation to eliminate $\dot{\phi}$ from the second of Lagrange's equations, resulting in a one-dimensional equation for the angle θ

$$\ddot{\theta} = \frac{L_z^2}{m^2 a^4} \frac{\cos\theta}{\sin^3\theta} - \frac{g}{a}\sin\theta$$

(b) Let us look for an equilibrium solution to this equation with $\theta = \theta_0 =$ constant, which yields

$$\frac{L_z^2}{m^2 a^4} = \frac{g\sin^4\theta_0}{a\cos\theta_0} \qquad ; \theta = \theta_0$$

This reproduces our previous result for the equilibrium circular orbits

$$\dot{\phi}^2 = \frac{g}{a\cos\theta_0}$$

(c) Suppose the system now suffers a perturbation that does not change the angular momentum L_z (the rotating particle is given a small transverse kick, for example). Let us look for small oscillations of the angle θ about the equilibrium value θ_0, at a fixed value of the angular momentum L_z

$$\theta = \theta_0 + \eta \qquad ; |\eta| \ll 1$$

Substitution into the above equation for $\ddot{\theta}$, and expansion for small η, give

$$\ddot{\eta} = \frac{L_z^2}{m^2 a^4} \frac{\cos(\theta_0 + \eta)}{\sin^3(\theta_0 + \eta)} - \frac{g}{a}\sin(\theta_0 + \eta)$$

$$\approx \frac{L_z^2}{m^2 a^4} \frac{[\cos\theta_0 - \eta\sin\theta_0]}{[\sin\theta_0 + \eta\cos\theta_0]^3} - \frac{g}{a}[\sin\theta_0 + \eta\cos\theta_0]$$

$$= \frac{L_z^2}{m^2 a^4} \frac{\cos\theta_0}{\sin^3\theta_0} \frac{[1 - \eta\tan\theta_0]}{[1 + \eta\cot\theta_0]^3} - \frac{g}{a}\sin\theta_0[1 + \eta\cot\theta_0]$$

Substitution of the equilibrium value of the first factor, and further expansion, give

$$\ddot{\eta} \approx \frac{g}{a}\sin\theta_0\left[1 - \eta\tan\theta_0 - 3\eta\cot\theta_0 - 1 - \eta\cot\theta_0\right]$$

$$= -\eta\frac{g}{a}\left[\frac{1 + 3\cos^2\theta_0}{\cos\theta_0}\right]$$

This is of the form

$$\ddot{\eta} = -\Omega^2\eta \qquad ; \Omega^2 = \frac{g}{a}\left[\frac{1 + 3\cos^2\theta_0}{\cos\theta_0}\right]$$

The orbits are stable, and the angle θ performs simple harmonic motion about the equilibrium value θ_0 as the particle goes around the circular orbit.

Problem 11.4 Suppose the particle on the table in the system in Sec. 9.2 is performing uniform circular motion with radius a, and the hanging particle is pulled down slightly.

(a) Show the angular momentum L_z is unchanged;

(b) Write $r(t) = a - \eta(t)$ where $|\eta(t)|/a \ll 1$, and use the analysis in Prob. 11.3 to show that the motion of the hanging particle is that of a simple harmonic oscillator;

(c) Show the angular frequency of the oscillation is $\omega^2 = 3g/2a$.

Solution to Problem 11.4

We follow exactly the same procedure as in the previous problem for the mechanical system in Sec. 9.2, as illustrated in Fig. 9.2 in the text.

(a) No torque is exerted on the system as the hanging particle is pulled down, so the angular momentum L_z of the sliding mass is unchanged.

(b) From Eq. (9.23), the angular momentum about the z-axis of the rotating particle is a constant of the motion

$$L_z = mr^2 \dot\theta = \text{constant}$$

The equation of motion for the radial coordinate of the rotating mass on the table is from Eq. (9.19)

$$2m\ddot r = -mg + mr\,\dot\theta^{\,2} = -mg + \frac{L_z^2}{mr^3}$$

The equilibrium circular orbit with constant $r = a$ is then described by

$$mg = \frac{L_z^2}{ma^3}$$

(c) Look for small deviations from this. Write $r = a - \eta$, and then

$$-2m\ddot\eta = -mg + \frac{L_z^2}{ma^3(1 - \eta/a)^3} \qquad ; r = a - \eta$$

$$\approx -mg + mg\left(1 + 3\frac{\eta}{a}\right)$$

(d) Thus the motion is *stable*, with

$$\ddot\eta = -\Omega^2\,\eta \qquad\qquad ; \Omega^2 = \frac{3g}{2a}$$

This simple harmonic motion is reflected in the position z of the hanging mass, since

$$z(t) = (l - a) + \eta(t) \qquad ; \text{ hanging mass}$$

Problem 11.5 Consider a marble rolling without slipping down an incline plane. Go through the arguments in Sec. 11.1.2, and make use of Prob. 8.4. Show that Eq. (11.27) becomes

$$\ddot{X} = -\frac{5g}{7} \sin \alpha$$

Solution to Problem 11.5

The arguments in Sec. 11.1.2 for the cylinder are exactly the same as for the marble down to Eq. (11.26)

$$\left(M + \frac{I}{a^2}\right) \ddot{X} = -Mg \sin \alpha$$

For the marble, from Prob. 8.4,

$$I = \frac{2}{5} Ma^2$$

Hence

$$\ddot{X} = -\frac{5g}{7} \sin \alpha$$

Chapter 12

Hamiltonian Dynamics

Problem 12.1 (a) Start from the hamiltonian for the simple harmonic oscillator in Prob. 11.1. Write Hamilton's equations, and show they reproduce Lagrange's equation and Newton's laws.

(b) Show the hamiltonian is a constant of the motion and also the energy;

(c) Show the phase-space orbit is an ellipse

$$\left(\frac{p}{a}\right)^2 + \left(\frac{q}{b}\right)^2 = 1$$

What are (a, b)?

Solution to Problem 12.1

(a) The hamiltonian for the simple harmonic oscillator in Prob. 11.1 is

$$H(p, q) = \frac{p^2}{2m} + \frac{1}{2}\kappa q^2 \qquad ; x \equiv q$$

Hamilton's equations then read

$$\frac{\partial H}{\partial p} = \frac{p}{m} = \frac{dq}{dt}$$

$$\frac{\partial H}{\partial q} = \kappa q = -\frac{dp}{dt}$$

$$\frac{dH}{dt} = \frac{\partial H}{\partial t} = 0$$

Substitution of the time derivative of the first equation into the second gives

$$m\frac{d^2 q}{dt^2} = -\kappa q$$

which reproduces Lagrange's equation and Newton's law.

(b) From the first relation in part (a), H is the total energy, and from the last of the second set of relations in part (a), it is a constant of the motion.

(c) It follows that the orbit in phase space is given by

$$\frac{p^2}{2m} + \frac{1}{2}\kappa q^2 = E = \text{constant}$$

This is rewritten as

$$\frac{p^2}{2mE} + \frac{\kappa q^2}{2E} = 1$$

Hence the orbit is an ellipse

$$\left(\frac{p}{a}\right)^2 + \left(\frac{q}{b}\right)^2 = 1$$

with

$$a^2 = 2mE \qquad ; \ b^2 = \frac{2E}{\kappa}$$

Problem 12.2 (a) A free particle moves along the x-axis. Plot its trajectory in phase space;

(b) The particle is now subject to a uniform accelerating force along the x-axis. Plot its new phase-space trajectory.

Solution to Problem 12.2

(a) For a free particle of mass m moving along the x-axis with velocity v_0, one has

$$p = p_0 \qquad\qquad ; \text{constant}$$
$$x = x_0 + \frac{p_0}{m}t$$

This orbit is shown in Fig. 12.1 below.

(b) If there is a uniform accelerating force F along the x-axis, then

$$p = p_0 + Ft$$
$$x = x_0 + \frac{p_0}{m}t + \frac{F}{2m}t^2$$

This orbit is also shown in Fig. 12.1.

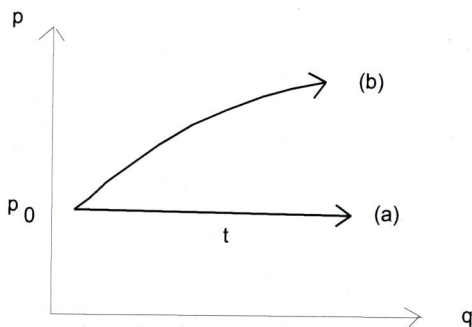

Fig. 12.1 Motion of a particle of mass m with time in phase space: (a) Free particle; (b) With accelerating force F along x-axis. Here $x \equiv q$.

Problem 12.3 Given the phase-space coordinate system (p, q), introduce the unit vectors $(\hat{p}, \hat{q})$, and write the differential displacements as

$$d\vec{p} = \hat{p}\, dp \qquad ; \; d\vec{q} = \hat{q}\, dq$$

(a) Show that the following vector product gives the area $\mathcal{A}(p, q)$ of a parallelogram with sides $(d\vec{p}, d\vec{q})$

$$d\vec{p} \times d\vec{q} = dp\, dq \sin\theta \, \hat{n} = \mathcal{A}(p, q)\, \hat{n}$$

where $\hat{n}$ is normal to the plane, and θ is the angle between $d\vec{p}$ and $d\vec{q}$;

(b) Go to another coordinate system where $p' = p'(p, q)$ and $q' = q'(p, q)$. Introduce the vector displacements[1]

$$d\vec{p}' = \left(\frac{\partial p'}{\partial p}\right) d\vec{p} + \left(\frac{\partial p'}{\partial q}\right) d\vec{q}$$

$$d\vec{q}' = \left(\frac{\partial q'}{\partial p}\right) d\vec{p} + \left(\frac{\partial q'}{\partial q}\right) d\vec{q}$$

Show

$$d\vec{p}' \times d\vec{q}' = \mathcal{A}(p', q')\, \hat{n}$$

$$= \left[\left(\frac{\partial p'}{\partial p}\right)\left(\frac{\partial q'}{\partial q}\right) - \left(\frac{\partial p'}{\partial q}\right)\left(\frac{\partial q'}{\partial p}\right) \right] \mathcal{A}(p, q)\, \hat{n}$$

---

[1]The transformation provides a mapping from the point (p, q) and infinitesimal vector displacements along the coordinate axes there of $(d\vec{p}, d\vec{q})$, to the new point (p', q') and corresponding displacements $(d\vec{p}', d\vec{q}')$. Expand $d\vec{p}' = a(p, q)\, d\vec{p} + b(p, q)\, d\vec{q}$. Now fix q. One then has $d\vec{p}' = dp'\hat{p} = a(p, q)\, dp\, \hat{p}$, where for fixed q [recall Eq. (2.26)], $dp' = (\partial p'/\partial p) dp$. It follows that $a(p, q) = \partial p'(p, q)/\partial p$, and so on.

(c) Hence, conclude that the differential areas are related by

$$\mathcal{A}(p', q') = \left| \frac{\partial(p', q')}{\partial(p, q)} \right| \mathcal{A}(p, q)$$

where the *jacobian determinant* is given by

$$\left| \frac{\partial(p', q')}{\partial(p, q)} \right| = \left(\frac{\partial p'}{\partial p} \right) \left(\frac{\partial q'}{\partial q} \right) - \left(\frac{\partial p'}{\partial q} \right) \left(\frac{\partial q'}{\partial p} \right)$$

$$; \text{ jacobian determinant}$$

(d) If both coordinate systems are *orthogonal,* conclude that the differential areas satisfy

$$dp' dq' = \left| \frac{\partial(p', q')}{\partial(p, q)} \right| dp dq$$

Solution to Problem 12.3

We are given a phase-space coordinate system (p, q) where the coordinates are linearly independent. At each point, we then have a set of coordinate axes with directions specified by unit vectors $(\hat{p}, \hat{q})$, where $\hat{p}$ points in the direction of increasing p at fixed q, and $\hat{q}$ points in the direction of increasing q at fixed p. We can then go a small distance along each axis with the vector displacements

$$d\vec{p} = \hat{p} \, dp \qquad ; \quad d\vec{q} = \hat{q} \, dq$$

(a) These vector displacements constitute the sides of a parallelogram. The following vector product gives the area $\mathcal{A}(p, q)$ of that parallelogram

$$d\vec{p} \times d\vec{q} = dp dq \sin \theta \, \hat{n} = \mathcal{A}(p, q) \, \hat{n}$$

Here $\hat{n}$ is normal to the plane, and θ is the angle between $d\vec{p}$ and $d\vec{q}$.

(b) Go to another coordinate system where $p' = p'(p, q)$ and $q' = q'(p, q)$. This transformation provides a mapping from the point (p, q) and infinitesimal vector displacements along the coordinate axes there of $(d\vec{p}, d\vec{q})$, to the new point (p', q') and corresponding vector displacements there of $(d\vec{p}', d\vec{q}')$. The displacement $d\vec{p}'$, which takes place at fixed q', can be expanded back in the basis $(d\vec{p}, d\vec{q})$ as

$$d\vec{p}' = a(p, q) \, d\vec{p} + b(p, q) \, d\vec{q}$$

Now fix q. One then has

$$dp\vec{\,}' = [dp']_q \, \hat{p} = a(p,q) \, dp \, \hat{p} \qquad \text{; fixed q}$$

It follows from Eq. (2.26) that

$$[dp']_q = \frac{\partial p'(p,q)}{\partial p} \, dp$$

Hence

$$a(p,q) = \frac{\partial p'(p,q)}{\partial p}$$

A similar argument gives $b(p,q) = \partial p'(p,q)/\partial q$. Thus we arrive at the expansion of the displacement $dp\vec{\,}'$ in terms of $(d\vec{p}, d\vec{q})$

$$d\vec{p}' = \left(\frac{\partial p'}{\partial p} \right) d\vec{p} + \left(\frac{\partial p'}{\partial q} \right) d\vec{q}$$

As a check, we can see that if we were to double the length of p' at fixed q', this expression would also double, and hence $d\vec{p}'$ has the right *direction*; its *length* has also been well-calculated from the above.[2]

In exactly the same fashion, we can express the displacement $d\vec{q}'$, which takes place at fixed p', in terms of $(d\vec{p}, d\vec{q})$

$$d\vec{q}' = \left(\frac{\partial q'}{\partial p} \right) d\vec{p} + \left(\frac{\partial q'}{\partial q} \right) d\vec{q}$$

It is now simple matter to compute the new phase-space area

$$d\vec{p}' \times d\vec{q}' = \mathcal{A}(p',q') \, \hat{n}$$
$$= \left[\left(\frac{\partial p'}{\partial p} \right) \left(\frac{\partial q'}{\partial q} \right) - \left(\frac{\partial p'}{\partial q} \right) \left(\frac{\partial q'}{\partial p} \right) \right] \mathcal{A}(p,q) \, \hat{n}$$

(c) Hence, we conclude that the differential areas are related by

$$\mathcal{A}(p',q') = \left| \frac{\partial(p',q')}{\partial(p,q)} \right| \mathcal{A}(p,q)$$

where the *jacobian determinant* is given by

$$\left| \frac{\partial(p',q')}{\partial(p,q)} \right| = \left(\frac{\partial p'}{\partial p} \right) \left(\frac{\partial q'}{\partial q} \right) - \left(\frac{\partial p'}{\partial q} \right) \left(\frac{\partial q'}{\partial p} \right)$$
$$\text{; jacobian determinant}$$

---

[2]Given dp', the quantities (dp, dq) are now determined.

(d) If both coordinate systems are *orthogonal*, we conclude that the differential areas satisfy

$$dp'\,dq' = \left| \frac{\partial(p',q')}{\partial(p,q)} \right| dp\,dq$$

Problem 12.4 In Prob. 12.3, let the original point in phase space (p,q) propagate after a time dt to a new point (p',q') through hamiltonian dynamics.

(a) Show that

$$p' = p + dp = p - \left(\frac{\partial H}{\partial q}\right) dt$$

$$q' = q + dq = q + \left(\frac{\partial H}{\partial p}\right) dt$$

(b) From which show that it follows

$$\frac{\partial p'}{\partial p} = 1 - \left(\frac{\partial^2 H}{\partial p \partial q}\right) dt \qquad ; \quad \frac{\partial p'}{\partial q} = -\left(\frac{\partial^2 H}{\partial q^2}\right) dt$$

$$\frac{\partial q'}{\partial q} = 1 + \left(\frac{\partial^2 H}{\partial q \partial p}\right) dt \qquad ; \quad \frac{\partial q'}{\partial p} = \left(\frac{\partial^2 H}{\partial p^2}\right) dt$$

(c) Therefore conclude that through order dt, the jacobian determinant is given by

$$\left| \frac{\partial(p',q')}{\partial(p,q)} \right| = 1 + \left[\left(\frac{\partial^2 H}{\partial q \partial p}\right) - \left(\frac{\partial^2 H}{\partial p \partial q}\right) \right] dt$$
$$= 1$$

(d) Hence, conclude that with hamiltonian dynamics, the differential phase-space area is *conserved*

$$\frac{d\mathcal{A}(p,q)}{dt} = 0 \qquad ; \text{ Liouville's theorem}$$

This is *Liouville's theorem*, which plays a central role in classical statistical mechanics.[3]

Solution to Problem 12.4

(a) For a system governed by hamiltonian dynamics, Hamilton's equations read

---

[3] See [Walecka (2000)].

$$\frac{dq}{dt} = \frac{\partial H}{\partial p} \qquad ; \qquad \frac{dp}{dt} = -\frac{\partial H}{\partial q}$$

Thus the new values of (p, q) after a short time interval dt are

$$p' = p + dp = p - \left(\frac{\partial H}{\partial q}\right) dt$$

$$q' = q + dq = q + \left(\frac{\partial H}{\partial p}\right) dt$$

(b) The following partial derivatives of $[p'(p, q), q'(p, q)]$ are readily calculated

$$\frac{\partial p'}{\partial p} = 1 - \left(\frac{\partial^2 H}{\partial p \partial q}\right) dt \qquad ; \qquad \frac{\partial p'}{\partial q} = -\left(\frac{\partial^2 H}{\partial q^2}\right) dt$$

$$\frac{\partial q'}{\partial q} = 1 + \left(\frac{\partial^2 H}{\partial q \partial p}\right) dt \qquad ; \qquad \frac{\partial q'}{\partial p} = \left(\frac{\partial^2 H}{\partial p^2}\right) dt$$

(c) The jacobian determinant is then given by

$$\left|\frac{\partial(p', q')}{\partial(p, q)}\right| = 1 + \left[\left(\frac{\partial^2 H}{\partial q \partial p}\right) - \left(\frac{\partial^2 H}{\partial p \partial q}\right)\right] dt + O\left[(dt)^2\right]$$

$$= 1 + O\left[(dt)^2\right]$$

(d) It follows from the previous problem that the differential phase-space areas are then related by

$$\mathcal{A}(p', q') = \mathcal{A}(p, q) + O\left[(dt)^2\right]$$

Hence, from the limit of $[\mathcal{A}(p', q') - \mathcal{A}(p, q)]/dt$,

$$\frac{d\mathcal{A}(p, q)}{dt} = 0 \qquad ; \text{ Liouville's theorem}$$

This is *Liouville's theorem*. The following problem provides an example.

Problem 12.5 Use an orthogonal coordinate system (p, q) to describe the motion of a free particle of mass m, and consider, at an initial time t_0, a finite rectangle with sides (a, b) parallel to the coordinate axes in that phase space.

(a) What is the hamiltonian $H(p, q)$ for this problem?

(b) Solve Hamilton's equations, and show what happens to that rectangle after a finite time interval $t - t_0$;[4]

(c) Show the area of that finite rectangle is preserved in time.

---

[4] *Hint:* Recall Eqs. (17.151).

Solution to Problem 12.5

(a) The hamiltonian for a free particle is

$$H(p,q) = \frac{p^2}{2m}$$

(b) Hamilton's equations read

$$\frac{dq}{dt} = \frac{\partial H}{\partial p} = \frac{p}{m}$$

$$\frac{dp}{dt} = -\frac{\partial H}{\partial q} = 0$$

The solution to these equations is

$$q = q_0 + \frac{p}{m}(t - t_0) \qquad ; p = \text{constant}$$

Figure 12.2 below shows a finite rectangle with sides (a, b) parallel to the orthogonal coordinate axes (p, q), at an initial time t_0, in the phase space used to describe the motion of that free particle. Also shown is what happens to that rectangle after a time $(t - t_0)$ according to Hamilton's equations. The height $a = p_2 - p_1$ remains unchanged since p is constant, as does the length of the base $b = q_2 - q_1$, again, since p is constant.

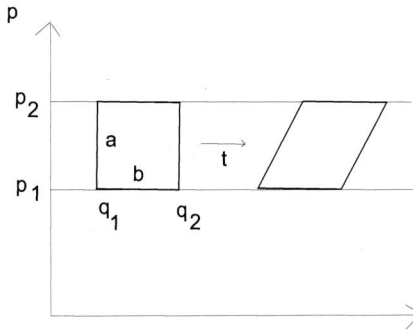

Fig. 12.2   A finite region in orthogonal phase space for free-particle motion.

(c) The initial rectangle is *skewed* with time; however, the *area* of the parallelogram, which is the product of the base times the height, remains *unchanged*

$$\mathcal{A} = ab \qquad ; \text{unchanged area}$$

Chapter 13

Continuum Mechanics

Problem 13.1 Suppose the springs in Fig. 13.1 in the text are fully extended in equilibrium so that they provide the tension τ in Eq. (13.17). Write each spring contribution as

$$\frac{\kappa}{2} \left[\mu(x_{i+1}) + a - \mu(x_i) \right]^2$$

(a) Construct the new lagrangian L;
(b) Show that Lagrange's equations are *unchanged*;
(c) Discuss the role of this modification in the lagrangian mechanics.

Solution to Problem 13.1

(a) The lagrangian in Eq. (13.3) with the fully extended springs becomes

$$L[\dot\mu(x_1), \cdots, \dot\mu(x_N), \mu(x_1), \cdots, \mu(x_N)] =$$

$$\frac{m}{2} \sum_{i=1}^{N} \dot\mu(x_i)^2 - \frac{\kappa}{2} \sum_{i=1}^{N} \left[\mu(x_{i+1}) + a - \mu(x_i) \right]^2$$

(b) Lagrange's Eqs. (13.4) now read

$$m\ddot\mu(x_j) = \kappa[\mu(x_{j+1}) + a - \mu(x_j)] - \kappa[\mu(x_j) + a - \mu(x_{j-1})]$$

$$; \; j = 1, 2, \cdots, N$$

The term in *a cancels and Lagrange's equations are unchanged*

$$m\ddot\mu(x_j) = \kappa[\mu(x_{j+1}) - \mu(x_j)] - \kappa[\mu(x_j) - \mu(x_{j-1})]$$

$$; \; j = 1, 2, \cdots, N$$

85

The force on each little mass comes from the difference of the extension of the neighboring springs, and the equilibrium length a cancels in that difference.

(c) The inclusion of the displacement a changes the equilibrium potential energy of each spring, so that the total equilibrium potential energy with vanishing displacements $\mu(x_j)$ is

$$V_0 = \frac{\kappa}{2} N a^2$$

Lagrangian mechanics, however, is independent of the choice of the zero of the potential.

Problem 13.2 It is also possible to understand the cutoff to the normal-mode frequencies $\omega_n(k_n)$ in Eq. (13.13), where

$$ka \le \pi \qquad ; n = 0, 1, \cdots, \frac{N}{2}$$

Make an argument that in the discrete system, it does not make sense to talk about waves with a half wavelength less than (approximately) the interparticle spacing. Hence

$$\lambda \ge 2a$$

Show this gives the above cutoff.

Solution to Problem 13.2

In a system with a continuous mass density, one can have longitudinal density waves with any wavelength propagating through it.

Where there is an underlying basis of discrete particles forming the mass distribution, one cannot have mass-density variations at a distance where there are no particles there to provide it! Hence there is a *cutoff* to the wavelength of the longitudinal density waves that can propagate through a discrete system. This cutoff will come when the wavelength is (approximately) twice the inter-particle spacing

$$\lambda_{\min} \approx 2a$$

This gives

$$k_{\max} = \frac{2\pi}{\lambda_{\min}} = \frac{\pi}{a}$$

Problem 13.3 (a) Show the normal-mode *eigenfunctions* in Sec. 13.3 satisfy the orthonormality relation

$$\frac{2}{l} \int_0^l dx \, \sin\left(\frac{n\pi x}{l}\right) \sin\left(\frac{m\pi x}{l}\right) = \delta_{m,n}$$

where $\delta_{m,n}$ is the Kronecker delta

$$\delta_{m,n} = 1 \quad ; \text{ if } m = n$$
$$= 0 \quad ; \text{ if } m \neq n$$

(b) Use this to construct the general solution in Eq. (13.32) for the string with fixed endpoints corresponding to the *initial conditions* [1]

$$q(x,0) = f(x)$$
$$\left[\frac{\partial q(x,t)}{\partial t}\right]_{t=0} = g(x)$$

Solution to Problem 13.3

(a) Consider the integral $I(m,n)$

$$I(m,n) \equiv \frac{2}{l} \int_0^l dx \, \sin\left(\frac{n\pi x}{l}\right) \sin\left(\frac{m\pi x}{l}\right)$$

where (m,n) are two positive integers. Use

$$\sin(a)\sin(b) = \frac{1}{2}\left[\cos(a-b) - \cos(a+b)\right]$$

Hence

$$I(m,n) = \frac{1}{l} \int_0^l dx \left\{\cos\left[\frac{(n-m)\pi x}{l}\right] - \cos\left[\frac{(n+m)\pi x}{l}\right]\right\}$$

Do the integrals

$$I(m,n) = \left\{\frac{1}{(n-m)\pi} \sin\left[\frac{(n-m)\pi x}{l}\right] - \frac{1}{(n+m)\pi} \sin\left[\frac{(n+m)\pi x}{l}\right]\right\}_0^l$$

Now observe

- If $m \neq n$, this expression vanishes;

---

[1] The wave equation is second-order in the time. Thus we must specify the function and its first time derivative everywhere, at the initial time, to define the solution. [Note we here correct a misprint in the last of Eqs. (13.33) in the text.]

- If $m = n$, the integral is 1.[2]

It follows that

$$I(m, n) = \delta_{m,n}$$

where $\delta_{m,n}$ is the Kronecker delta.

(b) The general solution for the motion of the string with fixed endpoints is

$$q(x,t) = \sum_{n=1}^{\infty} A_n \sin\left(\frac{n\pi x}{l}\right) \cos\left[\left(\frac{\pi n c t}{l}\right) + \chi_n\right] \quad ; \text{ general solution}$$

where the amplitudes and phases of the normal mode (A_n, χ_n) are constants to be determined from the initial conditions. As stated, we must specify the function and its first time derivative everywhere, at the initial time, to define the solution.

$$f(x) = \sum_{n=1}^{\infty} A_n \sin\left(\frac{n\pi x}{l}\right) \cos\left(\chi_n\right)$$

$$g(x) = -\sum_{n=1}^{\infty} \left(\frac{\pi n c}{l}\right) A_n \sin\left(\frac{n\pi x}{l}\right) \sin\left(\chi_n\right)$$

Now make use of the orthonormality of the eigenfunctions established in part (a)

$$\frac{2}{l} \int_0^l dx\, f(x) \sin\left(\frac{m\pi x}{l}\right) = A_m \cos\left(\chi_m\right)$$

$$\frac{2}{l} \int_0^l dx\, g(x) \sin\left(\frac{m\pi x}{l}\right) = -\left(\frac{\pi m c}{l}\right) A_m \sin\left(\chi_m\right)$$

This solves for (A_m, χ_m) for all m.[3]

Problem 13.4 Let the y-axis in Fig. 13.2 in the text point up in the lab. Suppose that while the end of the string at $x = 0$ is held fixed, the other end at $x = l$ is attached to a bead of mass m, which is free to move up and down without friction on a wire oriented in the vertical direction. The string is still under a uniform tension τ and undergoing planer motion. Neglect gravity.

---

[2]Start from $\sin^2(a) = [1 - \cos(2a)]/2$.
[3]Take the ratio for $\tan(\chi_m)$, and then use either one for A_m.

(a) Show that the equation of motion of the bead (for small displacements) is[4]

$$m\left(\frac{\partial^2 q}{\partial t^2}\right)_{x=l} = -\tau\left(\frac{\partial q}{\partial x}\right)_{x=l}$$

(b) Hence, show that as $m \to 0$, the boundary conditions for the string with *one free end* are

$$q(0,t) = 0$$
$$\left[\frac{\partial q(x,t)}{\partial x}\right]_{x=l} = 0$$

The slope at the free end must vanish.

Solution to Problem 13.4

(a) Concentrate on the bead on the vertical wire. Let the angle of the slope of the string attached to the bead be θ, and for small displacements $\sin\theta \approx \tan\theta$. If the slope is positive, the tension τ pulls the bead down, while if it is negative, it pulls it up. The force the tension in the string applies to the bead in the y-direction is thus

$$F_y = -\tau\sin\theta \approx -\tau\tan\theta = -\tau\left[\frac{\partial q}{\partial x}\right]_{x=l}$$

Newton's second law for motion of the bead in the y-direction then reads

$$m\left(\frac{\partial^2 q}{\partial t^2}\right)_{x=l} = -\tau\left(\frac{\partial q}{\partial x}\right)_{x=l}$$

(b) Now suppose the mass of the bead goes to zero while the tension of the string attached to the wire remains fixed at τ. This describes the case of a *free end*. The above then says that the slope of the string at this free end must *vanish*. The boundary conditions for the string with one fixed end and *one free end* are then

$$q(0,t) = 0$$
$$\left[\frac{\partial q(x,t)}{\partial x}\right]_{x=l} = 0 \qquad ;\text{ free end}$$

---

[4]There is a minus sign missing in this equation in the text; as written there, the relation would be appropriate if the new boundary conditions were at the origin.

Problem 13.5 Show that the normal-mode eigenfunctions and eigenvalues with the boundary conditions of Prob. 13.4 are

$$q_n(x) = \left(\frac{2}{l}\right)^{1/2} \sin\left[\frac{(2n+1)\pi x}{2l}\right] \qquad ; n = 0, 1, 2, \cdots, \infty$$

$$k_n = \frac{\omega_n}{c} = \frac{(2n+1)\pi}{2l}$$

Solution to Problem 13.5

Let us look for normal-mode solutions to the wave equation in Prob. 13.4 of the form

$$q(x,t) = q_n(x) \cos(\omega_n t + \chi_n)$$

The spatial boundary conditions are

$$q_n(0) = 0 \qquad ; \left[\frac{dq_n(x)}{dx}\right]_{x=l} = 0$$

The (normalized) normal-mode eigenfunctions and eigenvalues satisfying these boundary conditions are

$$q_n(x) = \left(\frac{2}{l}\right)^{1/2} \sin\left[\frac{(2n+1)\pi x}{2l}\right] \qquad ; n = 0, 1, 2, \cdots, \infty$$

$$k_n = \frac{\omega_n}{c} = \frac{(2n+1)\pi}{2l}$$

Problem 13.6 Suppose *both* ends of the string are free, in the sense of Prob. 13.4. Show that the normal-mode eigenfunctions and eigenvalues then become

$$q_n(x) = \left(\frac{2}{l}\right)^{1/2} \cos\left(\frac{n\pi x}{l}\right) \qquad ; n = 0, 1, 2, \cdots, \infty$$

$$k_n = \frac{\omega_n}{c} = \frac{n\pi}{l}$$

Note that with the exception of the $n = 0$ mode (describe that mode!),[5] this is the same spectrum as for the string with fixed ends in Eq. (13.30).

Solution to Problem 13.6

Now, instead of satisfying the boundary condition that both ends are fixed, the boundary condition is that the *derivatives* varnish at both ends,

---

[5]It is here unnormalized.

and the slopes are flat. The eigenfunctions, instead of being sines, are cosines, and the eigenfunctions and eigenvalues evidently are

$$q_n(x) = \left(\frac{2}{l}\right)^{1/2} \cos\left(\frac{n\pi x}{l}\right) \qquad ; n = 0, 1, 2, \cdots, \infty$$

$$k_n = \frac{\omega_n}{c} = \frac{n\pi}{l}$$

The solution with $n = 0$ corresponds to the flat string, under tension, sliding up and down the wires. It usually has to be dealt with on an individual basis.

Problem 13.7 The normal-mode spectrum for the discrete one-dimensional problem of masses coupled with springs is given in Eqs. (13.9) and (13.12). In the continuum limit, with the same periodic boundary conditions, the spectrum is given by

$$\omega^2 = (kc)^2$$

$$kl = 2n\pi \qquad ; n = 0, \pm 1, \pm 2, \cdots$$

Make a plot comparing these two spectra $\omega(k)$. [Recall Eqs. (13.19)–(13.20)].

Solution to Problem 13.7

Equations (13.12), (13.13), and (13.19) for the normal-mode spectrum for the discrete one-dimensional problem of masses coupled with springs can be written

$$\frac{\omega_n^2}{c^2} = 4\frac{\kappa\sigma}{m\tau} \sin^2\left(\frac{n\pi}{N}\right) \qquad ; c^2 = \frac{\tau}{\sigma}$$

With the aid of Eqs. (13.19) and (13.12) this becomes

$$\frac{\omega_n^2}{c^2} = \frac{4}{a^2} \sin^2\left(\frac{n\pi}{N}\right)$$

$$= k^2 \left[\left(\frac{N}{n\pi}\right)^2 \sin^2\left(\frac{n\pi}{N}\right)\right]$$

Hence, for the *discrete* system

$$\omega_n^2 = (kc)^2 \left[\left(\frac{N}{n\pi}\right)^2 \sin^2\left(\frac{n\pi}{N}\right)\right] \qquad ; n = 0, \pm 1, \pm 2, \cdots, \pm\frac{N}{2}$$

$$k = \frac{2n\pi}{Na}$$

while for the *continuous* system

$$\omega^2 = (kc)^2$$

$$k = \frac{2n\pi}{l} \qquad ; n = 0, \pm 1, \pm 2, \cdots$$

We compare the two dispersion relations for $\omega(k)$ in Fig. 13.1 below.

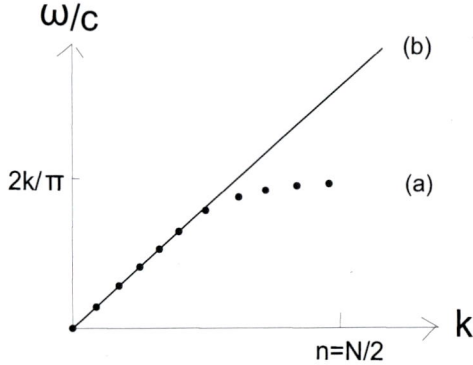

Fig. 13.1   Sketch of the normal-mode spectrum for longitudinal oscillations of masses coupled by springs; (a) Discrete mechanical system; (b) Continuum limit.

While the spectrum in the continuum limit continues on indefinitely with arbitrarily short wavelength,[6] in the case of the discrete mechanical system the spectrum levels off and terminates with a finite number of normal modes.[7]

The above figure, has $k \equiv |k|$. With periodic boundary conditions, there are horizontally reflected curves that describe waves traveling in the opposite direction, which we here simply suppress.

****

(*Aside*) As in Sec. 13.3, it is instructive to explicitly construct the continuum limit of the longitudinal oscillations of the mechanical system illustrated in Fig. 13.1 in the text.

Consider a long-wavelength normal mode where the displacements $\mu(x_i)$ become a continuous function $q(x)$. Now focus on a small element of the

---

[6]Recall $k = 2\pi/\lambda$.

[7]It should be pointed out that the continuum limit *also* consists of a set of points, with the same spacing in k if $Na = l$ (from the periodic boundary conditions), but these points lie on the indicated straight line that extends out to infinity with unit slope.

system of length Δl with n masses m, so that

$$\Delta l = na$$
$$\Delta m = nm = na(m/a) = \sigma \Delta l$$

where σ is the mass density

$$\sigma \equiv m/a \qquad \qquad ; \text{ mass density}$$

With fully stretched springs as in Prob. 13.1, the forces on the left and right masses of this element coming from the neighboring springs outside the element are then

$$F_l(x) = -\kappa\,[a + q(x) - q(x-a)] \approx -\kappa a \left\{ 1 + \left[\frac{\partial q}{\partial x}\right]_x \right\}$$

$$F_r(x + \Delta l) = \kappa\,[a + q(x + \Delta l + a) - q(x + \Delta l)] \approx \kappa a \left\{ 1 + \left[\frac{\partial q}{\partial x}\right]_{x+\Delta l} \right\}$$

Define the tension as

$$\tau \equiv \kappa a \qquad \qquad ; \text{ tension}$$

Then the net force on the little element is[8]

$$F_l(x) + F_r(x + \Delta l) \approx \tau \left\{ \left[\frac{\partial q}{\partial x}\right]_{x+\Delta l} - \left[\frac{\partial q}{\partial x}\right]_x \right\} \approx \tau \Delta l \frac{\partial^2 q}{\partial x^2}$$

Newton's second law for the little element then gives

$$\tau \Delta l \frac{\partial^2 q}{\partial x^2} = \sigma \Delta l \frac{\partial^2 q}{\partial t^2} \qquad \qquad ; \text{ Newton's law}$$

In the limit $\Delta l \to 0$, this is the one-dimensional wave equation for longitudinal density waves propagating around the mechanical system in Fig. 13.1 in the text

$$\frac{\partial^2 q(x,t)}{\partial x^2} = \frac{1}{c^2}\frac{\partial^2 q(x,t)}{\partial t^2} \qquad \qquad ; \ c^2 = \frac{\tau}{\sigma}$$

As one observable, it follows from the above that in leading order the mass density variation is given by

$$\frac{1}{\sigma_0}\delta\sigma(x,t) = -\frac{\partial q(x,t)}{\partial x} \qquad \qquad ; \text{ mass density}$$

---

[8]Compare the solution to Prob. 13.1(b). Notice that the "1" term from the fully extended springs again *cancels* in the net force on the element.

The oscillations of an elastic medium are discussed in the last chapter in [Fetter and Walecka (2003)]. In general, for each $\vec{k}$, there are two transverse shear waves and one longitudinal density wave with distinct velocities (c_t, c_l).

****

Chapter 14

Waves

Problem 14.1 A one-dimensional traveling wave moving in the x-direction is prepared in a non-dispersive medium where it satisfies the linear wave equation (a small amplitude displacement on a string under constant tension, for example). At the initial time $t = 0$ it has the shape

$$f(x) = Ae^{-x^2/a^2}$$

Plot it, and discuss what it looks like at a later time t. What might you expect it to look like in a medium where there is dispersion (waves on water, for example).

Solution to Problem 14.1

The aim of this problem is simply to have you get some familiarity with a non-dispersive wave, and we will be content here to just refer to Fig. 14.1 in the text. With no dispersion, as is the case for a small amplitude wave on a string, the disturbance propagates without change in shape, and the above function gives something pretty close to that in Fig. 14.1. You should try it yourself, sending a pulse down a long stretched string.

"No dispersion" means that the angular frequency and wavenumber are related by $\omega = kc$, where the velocity c is constant, so that the wave takes the form

$$\cos{(kx - \omega t)} = \cos{[k(x - ct)]}$$

There are many examples in nature where there is dispersion in the wave, and the pulse spreads out with time, and develops ripples. Throw a stone into a pond, for example, and see what happens. In this case, there is a more complicated relation between the angular frequency $\omega(k)$ and the wavenumber.[1]

Problem 14.2 Construct the standing-wave solutions to the wave equation obeying the spatial boundary conditions in Prob. 13.4.[2]

Solution to Problem 14.2

The boundary conditions in Prob. 13.4 describe a string under constant tension with one fixed and one free end. The appropriate eigenfunctions and eigenvalues are obtained in Prob. 13.5. The general solution to the wave equation is then obtained exactly as in Sec. 13.3.3

$$q(x,t) = \sum_{n=0}^{\infty} A_n \left(\frac{2}{l}\right)^{1/2} \sin\left[\frac{(2n+1)\pi x}{2l}\right] \cos\left\{\left[\frac{(2n+1)\pi ct}{2l}\right] + \chi_n\right\}$$

The coefficients (A_n, χ_n) are obtained from the initial conditions and the orthonormality of the eigenfunctions exactly as in Prob. 13.3.

Problem 14.3 Suppose $f(x)$ represents a plane-wave disturbance on a two-dimensional surface, perhaps a membrane, where the disturbance is independent of y at a given x. Let $\vec{x} = x\,\hat{x} + y\,\hat{y}$ denote an arbitrary position on that surface and let $\hat{n}$ be a unit-vector specifying some direction.

(a) Show $f(\vec{x} \cdot \hat{n} - ct)$ then describes a disturbance that moves without change in shape in the direction $\hat{n}$ with velocity c;

(b) Show this f obeys the *two-dimensional* wave equation

$$\left(\frac{\partial^2}{\partial x^2} + \frac{\partial^2}{\partial y^2}\right) f = \frac{1}{c^2}\frac{\partial^2 f}{\partial t^2} \qquad ; \text{two-dimensions}$$

Solution to Problem 14.3

(a) Suppose one simply rotates the underlying two-dimensional cartesian coordinate system so that the x-axis lies along $\hat{n}$. Then $f(\vec{x} \cdot \hat{n} - ct) = f(x - ct)$, and the discussion of the propagation of the wave is exactly the same as in Sec. 14.1 and Fig. 14.1 in the text.

(b) With the unrotated coordinate system

$$f(\vec{x} \cdot \hat{n} - ct) = f(xn_x + yn_y - t)$$

---

[1]As there is for the wave describing a free particle in quantum mechanics (see, for example, [Walecka (2008)]).

[2]Make use of Prob. 13.5.

Consider the following spatial derivatives [3]

$$\frac{\partial f}{\partial x} = n_x f' \qquad ; \qquad \frac{\partial f}{\partial y} = n_y f'$$

$$\frac{\partial^2 f}{\partial x^2} = n_x^2 f'' \qquad ; \qquad \frac{\partial^2 f}{\partial y^2} = n_y^2 f''$$

Since $\hat{n}$ is a unit vector

$$n_x^2 + n_y^2 = 1$$

The time derivatives are the same as in Sec. 14.1

$$\frac{1}{c^2} \frac{\partial^2 f}{\partial t^2} = f''$$

Hence this f obeys the *two-dimensional* wave equation

$$\left(\frac{\partial^2}{\partial x^2} + \frac{\partial^2}{\partial y^2} \right) f = \frac{1}{c^2} \frac{\partial^2 f}{\partial t^2} \qquad ; \text{ two-dimensions}$$

Problem 14.4 Consider a square membrane, clamped on all four sides of length l, obeying the two-dimensional wave equation of the previous problem.

(a) Look for separated, standing-wave solutions of the form

$$f(x, y, t) = A \sin (k_x x) \sin (k_y y) \cos (\omega t)$$

Show

$$k^2 = k_x^2 + k_y^2 = \frac{\omega^2}{c^2}$$

$$k_x = \frac{n_x \pi}{l} \qquad ; \ n_x = 1, 2, \cdots$$

$$k_y = \frac{n_y \pi}{l} \qquad ; \ n_y = 1, 2, \cdots$$

(b) Construct the general solution from the normal modes as in Eq. (13.32).

Solution to Problem 14.4

(a) Substitute this solution into the two-dimensional wave equation in Prob. 14.3

$$-(k_x^2 + k_y^2) f(x, y, t) = -\frac{\omega^2}{c^2} f(x, y, t)$$

---

[3] Here $f'(u) = df(u)/du$.

Hence, these normal modes must satisfy the dispersion relation

$$\omega^2 = (k_x^2 + k_y^2)c^2$$

They must also satisfy the spatial boundary condition of clamped edges in each direction, which gives (see Sec. 13.3.2)

$$k_x = \frac{m\pi}{l} \qquad ; m = 1, 2, \cdots$$

$$k_y = \frac{n\pi}{l} \qquad ; n = 1, 2, \cdots$$

Therefore, the normal-mode frequencies are given by

$$\omega_{mn}^2 = \left(\frac{m\pi c}{l}\right)^2 + \left(\frac{n\pi c}{l}\right)^2$$

(b) The general solution for the displacement $q(x, y, t)$ is obtained by superposing the normal modes

$$q(x, y, t) = \sum_{m=1}^{\infty} \sum_{n=1}^{\infty} A_{mn} \sin\left(\frac{m\pi x}{l}\right) \sin\left(\frac{n\pi y}{l}\right) \cos\left(\omega_{mn} t + \chi_{mn}\right)$$

The coefficients (A_{mn}, χ_{mn}) corresponding to a given set of initial conditions

$$q(x, y, 0) = f(x, y) \qquad ; \text{ initial conditions}$$

$$\left[\frac{\partial}{\partial t} q(x, y, t)\right]_{t=0} = g(x, y)$$

are then found exactly as in Prob. 13.3.

Chapter 15

Continuum Mechanics of String

Problem 15.1 Keep the next term in the expansion of Eq. (15.3), and obtain a *non-linear* correction to the wave equation for the string.[1]

Solution to Problem 15.1

With the aid of Prob. 10.3, the expansion of the square-root in the length ds of the small element of string in Eqs. (15.3)–(15.4) is extended to read

$$ds - dx \approx \frac{1}{2}\left[\frac{\partial q(x,t)}{\partial x}\right]^2 dx - \frac{1}{8}\left[\frac{\partial q(x,t)}{\partial x}\right]^4 dx$$

When multiplied by the tension τ, the potential energy of the displaced string is therefore incremented to

$$V = \int_0^l dx\, \mathcal{V}$$

$$\mathcal{V} = \frac{\tau}{2}\left[\frac{\partial q(x,t)}{\partial x}\right]^2 - \frac{\tau}{8}\left[\frac{\partial q(x,t)}{\partial x}\right]^4$$

The lagrangian density in Eq. (15.8) then becomes

$$\mathcal{L} = \frac{\sigma}{2}\left[\frac{\partial q(x,t)}{\partial t}\right]^2 - \frac{\tau}{2}\left[\frac{\partial q(x,t)}{\partial x}\right]^2 + \frac{\tau}{8}\left[\frac{\partial q(x,t)}{\partial x}\right]^4$$

$$\equiv \mathcal{L}_0 + \mathcal{L}_1$$

This lagrangian density still depends on the generalized coordinate $q(x,t)$ through $\mathcal{L}(\partial q/\partial t,\, \partial q/\partial x)$; however, Lagrange's equation now contains a

---

[1] Recall Prob. 10.3. Remember, in this chapter we are discussing the transverse planer oscillations of a string under tension.

nonlinear, cubic term in $\partial q(x,t)/\partial x$ arising from the additional quartic term $\mathcal{L}_1$ in the lagrangian density.

Problem 15.2 To get a feel for the momentum density, make a good sketch of $\mathcal{P}_\perp(x,t)$, $\vec{\mathcal{P}}(x,t)$, and $\mathcal{P}_x(x,t)$ at several points for the traveling wave in Eq. (15.46).

Solution to Problem 15.2

The traveling wave on a string in Eq. (15.46) is

$$\frac{1}{A}q(x,t) = \cos\left[k(x - ct)\right] \qquad ; \text{ traveling wave}$$

The x-component of the momentum density in this wave is given in Eq. (15.53)

$$\frac{1}{c\sigma(kA)^2}\mathcal{P}_x(x,t) = \sin^2\left[k(x - ct)\right] \qquad ; \text{ momentum density}$$

In order to get a feel for the momentum density, instead of looking at the individual components, we just compare these two quantities in Fig. 15.1 below.

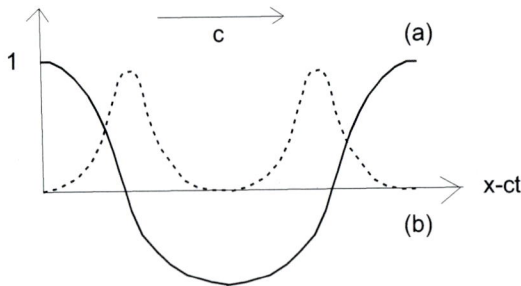

Fig. 15.1   Sketch of momentum density in direction of motion for traveling wave on a string: (a) Traveling wave $q(x,t)/A$; and (b) Momentum density $\mathcal{P}_x(x,t)/c\sigma(kA)^2$.

The momentum density in the wave in the direction of motion moves with the wave velocity c, and within one wavelength, it comes in *two* pulses where the magnitude of the slope of the displacement is maximized. This momentum density vanishes where that slope vanishes, and in this region of the wave, there is no momentum transport.

For the string, the momentum flux is related to the energy flux by Eq. (15.55)

$$c\, \mathcal{P}_x(x,t) = \frac{1}{c} S_x(x,t) \qquad ; \text{ momentum flux}$$

Note that $\mathcal{P}_x(x,t)$ and $S_x(x,t)$ have the *same shape*.

Problem 15.3 Suppose that the mass density $\sigma(x)$ and tension $\tau(x)$ are both specified functions of position in the lagrangian density for the string in Eq. (15.29). Show the wave equation becomes

$$\frac{\partial^2 q(x,t)}{\partial t^2} = \frac{1}{\sigma(x)} \frac{\partial}{\partial x} \left[\tau(x) \frac{\partial q(x,t)}{\partial x} \right]$$

Solution to Problem 15.3

The lagrangian density is now

$$\mathcal{L} = \frac{\sigma(x)}{2} \left[\frac{\partial q(x,t)}{\partial t} \right]^2 - \frac{\tau(x)}{2} \left[\frac{\partial q(x,t)}{\partial x} \right]^2$$

Lagrange's equation reads

$$\frac{\partial}{\partial t} \left[\frac{\partial \mathcal{L}}{\partial(\partial q/\partial t)} \right] + \frac{\partial}{\partial x} \left[\frac{\partial \mathcal{L}}{\partial(\partial q/\partial x)} \right] - \frac{\partial \mathcal{L}}{\partial q} = 0$$

The required partial derivatives of the lagrangian density are given by

$$\frac{\partial \mathcal{L}}{\partial(\partial q/\partial t)} = \sigma(x) \frac{\partial q(x,t)}{\partial t}$$

$$\frac{\partial \mathcal{L}}{\partial(\partial q/\partial x)} = -\tau(x) \frac{\partial q(x,t)}{\partial x}$$

Hence, in this case Lagrange's equation reads

$$\frac{\partial^2 q(x,t)}{\partial t^2} = \frac{1}{\sigma(x)} \frac{\partial}{\partial x} \left[\tau(x) \frac{\partial q(x,t)}{\partial x} \right]$$

Problem 15.4 Consider a semi-infinite string with $x \geq 0$ fixed at the origin so that the boundary condition is

$$q(0,t) = 0$$

(a) Show that a normal-mode solution on the infinite string satisfying this boundary condition is

$$q(x,t) = A \left[\sin k(x + ct) + \sin k(x - ct) \right]$$

For $x \geq 0$, interpret this as an incident wave and a reflected wave

$$q(x,t) = q^{\text{inc}}(x,t) + q^{\text{refl}}(x,t)$$

(b) The energy flux is given in Eq. (15.42)

$$S_x(x,t) = -\tau \frac{\partial q(x,t)}{\partial x} \frac{\partial q(x,t)}{\partial t}$$

Show that the reflected energy flux is given by

$$S_x^{\text{refl}}(x,t) = \tau c(kA)^2 \cos^2 k(x - ct) \qquad ; x \geq 0$$

This energy flux is periodic and moves with the wave in the positive x-direction (see Sec. 14.1).

Show that the incident energy flux at a corresponding position on the wave front is given by[2]

$$S_x^{\text{inc}}(x,t) = -\tau c(kA)^2 \cos^2 k(-x + ct) \qquad ; x \geq 0$$

(c) Define a reflection coefficient $\mathcal{R}$ as the magnitude of the ratio of the reflected and incident energy fluxes. Show

$$\mathcal{R}(x) \equiv \left| \frac{S_x^{\text{refl}}(x,t)}{S_x^{\text{inc}}(x,t)} \right| = 1$$

Interpret this result.

Solution to Problem 15.4

We are looking at the semi-infinite, positive-x, string under constant tension τ, with a support at the origin that clamps the string and imposes the boundary condition

$$q(0,t) = 0 \qquad ; \text{boundary condition}$$

(a) A normal-mode solution to the wave equation on the string that will allow us to satisfy this boundary condition is the superposition

$$q(x,t) = A\left[\sin k(x + ct) + \sin k(x - ct)\right] \qquad ; \omega = kc$$

---

[2] An alternate definition of the incident flux is $S_x^{\text{inc}}(x,t) \equiv S_x^{\text{inc}}(x,t_{\text{inc}})$ where $t_{\text{inc}} = t - 2x/c$ is the time, retarded by the amount it takes for the sound wave to get to the wall and back.

One can interpret this as an *incident wave* moving to the left, and a *reflected wave* moving to the right (see Sec. 14.1)

$$q(x,t) = q^{\text{inc}}(x,t) + q^{\text{refl}}(x,t)$$

The boundary condition at the origin is satisfied by this wave

$$q(0,t) = A\left[\sin(\omega t) + \sin(-\omega t)\right] = A\left[\sin \omega t - \sin \omega t\right] = 0$$

At the support, the reflected wave has the same amplitude as the incident wave, but it has a phase change of π.

(b) The energy flux on the string in the x-direction is given by Eq. (15.42)

$$S_x(x,t) = -\tau \frac{\partial q(x,t)}{\partial x}\frac{\partial q(x,t)}{\partial t}$$

The energy flux of the *reflected wave* is then immediately calculated to be

$$S_x^{\text{refl}}(x,t) = \tau c(kA)^2 \cos^2 k(x-ct)$$

This energy flux is periodic and moves with the wave in the positive x-direction.

The simultaneous energy flux of the *incident wave* is given by

$$S_x^{\text{inc}}(x,t) = -\tau c(kA)^2 \cos^2 k(x+ct)$$

This incident energy flux is also periodic and moves with the wave in the negative x-direction.

(c) In order to compare the fluxes at a comparable position on the wavefront, let us evaluate the incident flux at a *retarted* time t_R, where $t_R = t - 2x/c$ is the time it takes for the sound wave to get to the wall and back

$$S_x^{\text{inc}}(x,t_R) = -\tau c(kA)^2 \cos^2 k(-x+ct) \qquad ; t_R = t - \frac{2x}{c}$$
$$= -\tau c(kA)^2 \cos^2 k(x-ct)$$

The energy fluxes of the incident and reflected waves are then equal and opposite, and the reflection coefficient $\mathcal{R}$ is unity

$$\mathcal{R}(x) \equiv \left| \frac{S_x^{\text{refl}}(x,t)}{S_x^{\text{inc}}(x,t_R)} \right| = 1$$

This is just energy conservation. The support does not move, and it does no work on the system. Hence the energy that the incident wave brings in is exactly the same as the energy the reflected wave later carries out.

Problem 15.5 Suppose the support at the origin to which the end of the semi-infinite string in Prob. 15.4 is fixed is oscillating with an angular frequency ω

$$q(0, t) = A \sin \omega t$$

(a) Show that the driven, outgoing-wave solution to the wave equation for the string that matches this boundary condition at the origin is

$$q^{\mathrm{rad}}(x, t) = -A \sin k(x - ct) \qquad ; \omega = kc \quad ; x \geq 0$$

(b) Show that the corresponding radiated energy flux in the string arising from the oscillating support is

$$S_x^{\mathrm{rad}}(x, t) = \tau c(kA)^2 \cos^2 k(x - ct)$$

Interpret this result.

Solution to Problem 15.5

(a) An outgoing-wave solution to the wave equation on the positive x-axis for the semi-infinite string is

$$q(x, t) \propto \sin k(x - ct) \qquad ; \text{ outgoing wave}$$

This must now be matched to the boundary condition of an oscillating support at the origin

$$q(0, t) = A \sin \omega t \qquad ; \text{ boundary condition}$$

The solution is clearly

$$q(x, 0) = -A \sin k(x - ct) \qquad ; kc = \omega$$

(b) The energy flux carried by this outgoing wave follows from Eqs. (15.42)

$$S_x(x, t) = -\tau \frac{\partial q(x, t)}{\partial x} \frac{\partial q(x, t)}{\partial t}$$

This gives

$$S_x^{\mathrm{rad}}(x, t) = \tau c(kA)^2 \cos^2 k(x - ct)$$

This is the basic phenomenon of *radiation*. The oscillating support puts energy into the system, here by doing work against the tension in the string. The oscillating support sends out a traveling wave that carries this energy. The energy can then be transported by the wave to another place, where the radiation signal can then be detected.

Problem 15.6 Suppose one attaches a mass m at the midpoint of the string with fixed ends examined in Sec. 13.3. Write Newton's second law for that mass.

(a) Show that the lowest frequency normal-mode eigenfunction (continuous and unnormalized) can be written as

$$q(x) = \sin kx \qquad ; x < \frac{l}{2}$$

$$q(x) = \sin k(l - x) \qquad ; x > \frac{l}{2}$$

(b) Show that the eigenfrequencies $\omega = kc$ satisfy the equation

$$\left(\frac{kl}{2} \right) \tan \left(\frac{kl}{2} \right) = \frac{\sigma l}{m} = \frac{m_{\text{string}}}{m}$$

(c) Show that if the attached mass is negligible, this reproduces the result in Sec. 13.3

$$k = \frac{\pi}{l} \qquad ; m \ll m_{\text{string}}$$

(d) Show that if m is very heavy it does not move, and there will be a pair of matching arcs in the first of Figs. 13.3 in the text with

$$k = \frac{2\pi}{l} \qquad ; m \gg m_{\text{string}}$$

Solution to Problem 15.6

(a) The normal-mode solutions to the wave equation take the form

$$q(x, t) = q(x) \cos (\omega t) \qquad ; \text{normal modes}$$

$$\frac{d^2 q(x)}{dx^2} = -k^2 q(x) \qquad ; k^2 \equiv \frac{\omega^2}{c^2}$$

The lowest frequency normal-mode eigenfunction (continuous and unnormalized), which vanishes at the two ends, can then be written as

$$q(x) = \sin kx \qquad\qquad ; \; x < \frac{l}{2}$$

$$q(x) = \sin k(l - x) \qquad ; \; x > \frac{l}{2}$$

(b) If a mass m is placed at the midpoint of the string, the tension in the string will pull that mass back toward the axis, and Newton's second law for that mass then reads

$$\tau \left\{ \left[-\frac{\partial q(x,t)}{\partial x} \right]_{x=l/2^-} + \left[\frac{\partial q(x,t)}{\partial x} \right]_{x=l/2^+} \right\} = m \frac{\partial^2 q(x,t)}{\partial t^2}$$

Substitution of the above normal-mode solution gives

$$-\tau \left[2k \cos \left(\frac{kl}{2} \right) \right] = -m\omega^2 \sin \left(\frac{kl}{2} \right)$$

Use $\omega^2 = k^2 c^2 = k^2 \tau / \sigma$, and this eigenvalue equation takes the form

$$\left(\frac{kl}{2} \right) \tan \left(\frac{kl}{2} \right) = \frac{\sigma l}{m} = \frac{m_{\text{string}}}{m}$$

(c) If the attached mass is tiny the r.h.s. becomes very large, and $kl/2 \to \pi/2$. Hence

$$k = \frac{\pi}{l} \qquad\qquad ; \; m \ll m_{\text{string}}$$

The eigenfunction then looks like that in the top of Fig. 13.3 in the text.

(d) If m is very heavy it does not move, the r.h.s. of the above vanishes, and $kl/2 \to \pi$. Thus

$$k = \frac{2\pi}{l} \qquad\qquad ; \; m \gg m_{\text{string}}$$

There will then be a pair of matching arcs in the first of Figs. 13.3.

Problem 15.7 Show the energy-momentum tensor for the string in Eq. (15.58) is *symmetric*, with

$$\mathcal{T}_{21}(x,t) = -ic\sigma \frac{\partial q(x,t)}{\partial t} \frac{\partial q(x,t)}{\partial x} = ic \, \mathcal{P}_x(x,t)$$

$$= \mathcal{T}_{12}(x,t)$$

Solution to Problem 15.7

The energy-momentum tensor for the string in Eq, (15.58) is

$$T_{\mu\nu} \equiv \mathcal{L}\delta_{\mu\nu} - \frac{\partial\mathcal{L}}{\partial(\partial q/\partial x_\mu)}\frac{\partial q}{\partial x_\nu}$$

where the lagrangian density is given in Eq. (15.8)

$$\mathcal{L} = \frac{\sigma}{2}\left[\frac{\partial q(x,t)}{\partial t}\right]^2 - \frac{\tau}{2}\left[\frac{\partial q(x,t)}{\partial x}\right]^2$$

It follows that

$$T_{21}(x,t) = -ic\sigma\frac{\partial q(x,t)}{\partial t}\frac{\partial q(x,t)}{\partial x} = ic\,\mathcal{P}_x(x,t)$$
$$= T_{12}(x,t)$$

Here $\mathcal{P}_x(x,t)$ is the momentum density in Eq. (15.52).[3] Thus the energy-momentum tensor for the string is symmetric.

Problem 15.8 (a) Show that the energy-momentum tensor for the string gives

$$T_{11}(x,t) = \mathcal{H}(x,t)$$

(b) Show the $\nu = 1$ component of the conservation law for the energy-momentum tensor in Eq. (15.68) gives a continuity equation for the momentum density

$$\frac{\partial}{\partial x_\mu}T_{\mu 1} = \frac{\partial\mathcal{P}_x(x,t)}{\partial t} + \frac{\partial\mathcal{H}(x,t)}{\partial x} = 0$$

Show this is satisfied by the traveling wave in Eq. (15.53);

(c) Suppose one has a string of length l satisfying periodic boundary conditions. Show that it follows from the result in part (b) that the total momentum in the x-direction is a constant of the motion

$$\frac{dP_x}{dt} = \frac{d}{dt}\int_0^l \mathcal{P}_x(x,t)dx = 0$$

---

[3]Compare the similar relation for a fluid in the *Aside* after Prob. 16.12.

Solution to Problem 15.8

(a) It follows from the previous problem that

$$T_{11}(x,t) = \frac{\sigma}{2}\left[\frac{\partial q(x,t)}{\partial t}\right]^2 + \frac{\tau}{2}\left[\frac{\partial q(x,t)}{\partial x}\right]^2$$
$$= \mathcal{H}(x,t)$$

where $\mathcal{H}(x,t)$ is the hamiltonian (energy) density of Eq. (15.35).

(b) If $\mathcal{L}$ has no *explicit* dependence on x_μ, then Eq. (15.68) says that the energy-momentum tensor is conserved

$$\frac{\partial}{\partial x_\mu}T_{\mu\nu} = 0$$

Write this out for $\nu = 1$ using the previous results

$$\frac{\partial}{\partial x_\mu}T_{\mu 1} = \frac{\partial \mathcal{P}_x(x,t)}{\partial t} + \frac{\partial \mathcal{H}(x,t)}{\partial x} = 0$$

This is the continuity equation for the momentum density $\mathcal{P}_x(x,t)$.

The traveling wave in Eq. (15.53) gives

$$\mathcal{P}_x(x,t) = c\sigma(kA)^2\sin^2\left[k(x-ct)\right] = \frac{1}{c}\mathcal{H}(x,t)$$

which satisfies the above continuity equation.

(c) Consider the time derivative of the integral of the momentum density over the length of the string, and use the continuity equation

$$\frac{d}{dt}\int_0^l dx\,\mathcal{P}_x(x,t) = \int_0^l dx\,\frac{\partial \mathcal{P}_x(x,t)}{\partial t} = -\int_0^l dx\,\frac{\partial \mathcal{H}(x,t)}{\partial x}$$
$$= -\left[\mathcal{H}(x,t)\right]_0^l$$

With periodic periodic boundary conditions this vanishes, and the total momentum in the x-direction is a constant of the motion.

Problem 15.9 Make use of Lagrange's equations to show that the energy-momentum tensor with several generalized coordinates in Eq. (15.72) satisfies the conservation law in Eq. (15.68).

Solution to Problem 15.9

We are given a lagrangian density with several generalized coordinates $\mathcal{L}(\partial q_1/\partial x_\mu, \cdots, \partial q_N/\partial x_\mu, q_1, \cdots, q_N; x_\mu)$, and the energy-momentum tensor of Eq. (15.72),

$$T_{\mu\nu} \equiv \mathcal{L}\delta_{\mu\nu} - \sum_{i=1}^{N} \frac{\partial \mathcal{L}}{\partial(\partial q_i/\partial x_\mu)} \frac{\partial q_i}{\partial x_\nu}$$

Let us go through the derivation of the conservation of that tensor in Sec. 15.6, making use of Lagrange's Eqs. (10.25),

$$\frac{d}{dt}\left(\frac{\partial L}{\partial \dot{q}_i}\right) - \frac{\partial L}{\partial q_i} = 0 \qquad ; i = 1, \cdots, N$$

In Eqs. (15.65) and (15.66), all but the last term in the latter equation now carry an additional $\sum_{i=1}^{N}$. All the observed cancellations again occur, and Eqs. (15.67)–(15.68) are again valid.

Hence, it is still true that if the lagrangian density has no *explicit* dependence on x_μ, then

$$\frac{\partial}{\partial x_\mu} T_{\mu\nu} = 0 \qquad ; \text{ given } \mathcal{L}\left(\frac{\partial q_i}{\partial x_\mu}, q_i\right)$$

Problem 15.10 (a) Use the notation of Sec. 15.6, and show that the lagrangian density for the string can be written in the following ("invariant") form

$$\mathcal{L} = -\frac{\tau}{2}\left(\frac{\partial q}{\partial x_\mu}\right)\left(\frac{\partial q}{\partial x_\mu}\right) = -\frac{\tau}{2}\left(\frac{\partial q}{\partial x_\mu}\right)^2 \qquad ; \text{ string}$$

(b) Show that Lagrange's equation can similarly be written as

$$\frac{\partial}{\partial x_\mu}\left[\frac{\partial \mathcal{L}}{\partial(\partial q/\partial x_\mu)}\right] - \frac{\partial \mathcal{L}}{\partial q} = 0 \qquad ; \text{ string}$$

Solution to Problem 15.10

(a) The lagrangian density for the string is

$$\mathcal{L} = \frac{\sigma}{2}\left[\frac{\partial q(x,t)}{\partial t}\right]^2 - \frac{\tau}{2}\left[\frac{\partial q(x,t)}{\partial x}\right]^2$$

Introduce the two-vector $x_\mu = (x_1, x_2) = (x, ict)$, with $c^2 = \tau/\sigma$ and i the imaginary number $\sqrt{-1}$. Also, introduce the convention that repeated Greek indices are summed from 1 to 2. This lagrangian density can then be written as

$$\mathcal{L} = -\frac{\tau}{2}\left(\frac{\partial q}{\partial x_\mu}\right)\left(\frac{\partial q}{\partial x_\mu}\right) = -\frac{\tau}{2}\left(\frac{\partial q}{\partial x_\mu}\right)^2 \qquad \text{; string}$$

(b) Lagrange's equation for the string is

$$\frac{\partial}{\partial t}\left[\frac{\partial \mathcal{L}}{\partial(\partial q/\partial t)}\right] + \frac{\partial}{\partial x}\left[\frac{\partial \mathcal{L}}{\partial(\partial q/\partial x)}\right] - \frac{\partial \mathcal{L}}{\partial q} = 0 \qquad \text{; Lagrange's eqn}$$

With the above conventions, this can be written as

$$\frac{\partial}{\partial x_\mu}\left[\frac{\partial \mathcal{L}}{\partial(\partial q/\partial x_\mu)}\right] - \frac{\partial \mathcal{L}}{\partial q} = 0 \qquad \text{; string}$$

Chapter 16

Mechanics of Fluids

Problem 16.1 (a) Consider the equation of *hydrostatic equilibrium* for a fluid at the earth's surface with a downward pointing force per unit mass of $\vec{f}_{ext} = \vec{g}$. Show that Newton's law in Eq. (16.36) gives

$$\vec{\nabla}P = \rho\vec{g} \qquad \text{; hydrostatics}$$

where P is the pressure and ρ is the mass density (assumed constant here);

(b) Let z point down from the earth's surface. Show the ocean pressure is

$$P(z) = P_0 + \rho_{\text{water}}\,gz \qquad \text{; ocean pressure}$$

Solution to Problem 16.1

(a) Newton's law for fluid flow in Eq. (16.36) reads

$$\frac{d\vec{v}}{dt} = -\frac{1}{\rho}\vec{\nabla}P + \vec{f}_{ext} \qquad \text{; Newton's second law}$$

where the force $\vec{f}_{ext}$ is per unit mass. If the fluid is *stationary*, then

$$\vec{\nabla}P = \rho\vec{f}_{ext} \qquad \text{; hydrostatics}$$

(b) With a gravitational force $\vec{f}_{grav} = \vec{g}$ pointing in the z-direction (down), and a one-dimensional spatial dependence, this relation reads

$$\frac{dP(z)}{dz} = \rho g$$

The solution to this equation, for the ocean with a constant ρ_{water}, is

$$P(z) = P_0 + \rho_{\text{water}}\,gz \qquad \text{; ocean pressure}$$

111

Problem 16.2 Use the following values for the speed of sound in air under standard conditions, and the frequency of middle C

$$c_{\text{sound}} = 331 \, \text{m/sec}$$

$$\nu_C = 261.6 \, /\text{sec}$$

Make a good numerical calculation of $v_x/(kA)$, $c\,\delta\rho/(kA)\rho_0$, and $S_x/c\rho_0(kA)^2$ for the sound wave in Eqs. (16.66)–(16.68).[1]

Solution to Problem 16.2

In chapter 16 we studied a sound wave with the velocity potential given in Eq. (16.64)

$$\Phi(\vec{x}, t) = \mathcal{A} \cos\left[k(x - ct)\right]$$

The fluid velocity, density variation, and energy flux for this wave are given in Eqs. (16.66)–(16.68)

$$\frac{1}{(k\mathcal{A})} v_x(\vec{x}, t) = \frac{c}{\rho_0(k\mathcal{A})} \delta\rho(\vec{x}, t) = \sin\left[k(x - ct)\right]$$

$$\frac{1}{c\rho_0(k\mathcal{A})^2} S_x(\vec{x}, t) = \sin^2\left[k(x - ct)\right]$$

The goal of this problem is to put some numbers into these relations. We have

$$\lambda = \frac{c}{\nu} = \frac{331 \, \text{m/sec}}{261.6 \, /\text{sec}} = 1.26 \, \text{m} \qquad ; \text{ sound}$$

$$k = \frac{2\pi}{\lambda} = 4.96 \, /\text{m}$$

$$\omega = kc = 1.64 \times 10^3 \, /\text{sec}$$

Problem 16.3 (a) Show that the vector field $\vec{v}(\vec{x}, t) = -\vec{\nabla}\Phi(\vec{x}, t)$ is *irrotational*[2]

$$\vec{\nabla} \times \vec{v}(\vec{x}, t) = -\vec{\nabla} \times \vec{\nabla}\Phi(\vec{x}, t) = 0$$

(b) Explicitly construct the gauge transformation that absorbs the constant term $(\epsilon + P/\rho)_0$ in Eq. (16.54) into Φ.

---

[1] The corresponding value of the mass density is $\rho_0 = 1.292 \, \text{kg/m}^3$. What is k?

[2] The partial derivatives here keep all the other members of the variable set (x, y, z, t) constant; they are the derivatives at any *instant*.

Solution to Problem 16.3

(a) The gradient is given by

$$\vec{\nabla} \equiv \hat{x}\frac{\partial}{\partial x} + \hat{y}\frac{\partial}{\partial y} + \hat{z}\frac{\partial}{\partial z} \qquad ; \text{ gradient}$$

The derivatives act on a following function, and the cartesian unit vectors are constants. The curl of the gradient is then given by

$$\text{curl}\,(\text{grad}\,\Phi) = \left(\vec{\nabla} \times \vec{\nabla}\right)\Phi$$

But from its definition, the cross product of two identical vectors vanishes, so that[3]

$$\left(\vec{\nabla} \times \vec{\nabla}\right)\Phi = 0$$

(b) The gauge transformation that absorbs the constant term $(\epsilon + P/\rho)_0$ in Eq. (16.54), which appears in Bernoulli's theorem in Eq. (16.49), is

$$\Phi \to \Phi + \left(\epsilon + \frac{P}{\rho}\right)_0 t$$

Problem 16.4 Make use of the lagrangian density in Eq. (16.16) and the definition in Eq. (16.2) to derive Eq. (16.20), which is used to obtain both the continuity equation and energy flux.

Solution to Problem 16.4

The lagrangian density in Eq. (16.16) is

$$\mathcal{L}\left(\frac{\partial \Phi}{\partial t}, \vec{\nabla}\Phi, \rho; \vec{x}\right) = \rho\frac{\partial \Phi}{\partial t} - \frac{1}{2}\rho(\vec{\nabla}\Phi)^2 - \rho\epsilon(\rho) - \rho U(\vec{x})$$
$$; \text{ lagrangian density for irrotational isentropic flow}$$

The definition in Eq. (16.2) is

$$\frac{\partial}{\partial x}\left[\frac{\partial \mathcal{L}}{\partial(\partial q/\partial x)}\right] \to \frac{\partial}{\partial x}\left[\frac{\partial \mathcal{L}}{\partial(\partial q/\partial x)}\right] + \frac{\partial}{\partial y}\left[\frac{\partial \mathcal{L}}{\partial(\partial q/\partial y)}\right] + \frac{\partial}{\partial z}\left[\frac{\partial \mathcal{L}}{\partial(\partial q/\partial z)}\right]$$
$$\equiv \vec{\nabla} \cdot \left[\frac{\partial \mathcal{L}}{\partial(\vec{\nabla}q)}\right]$$

---

[3] Recall that the partial derivatives can be taken in any order.

This has to be a definition since a gradient in the denominator by itself is undefined; however, it is clear from this relation that[4]

$$\frac{\partial \mathcal{L}}{\partial (\vec{\nabla} q)} \equiv \hat{x} \frac{\partial \mathcal{L}}{\partial (\partial q/\partial x)} + \hat{y} \frac{\partial \mathcal{L}}{\partial (\partial q/\partial y)} + \hat{z} \frac{\partial \mathcal{L}}{\partial (\partial q/\partial z)}$$

From the above lagrangian density

$$\frac{\partial \mathcal{L}}{\partial (\partial \Phi/\partial x)} = -\rho \frac{\partial \Phi}{\partial x}$$

Since this also holds for the other two components

$$\frac{\partial \mathcal{L}}{\partial (\vec{\nabla} \Phi)} = -\rho \vec{\nabla} \Phi$$

where

$$\vec{\nabla} \equiv \hat{x} \frac{\partial}{\partial x} + \hat{y} \frac{\partial}{\partial y} + \hat{z} \frac{\partial}{\partial z} \qquad ; \text{ gradient}$$

This is Eq. (16.20).

Problem 16.5 (a) Present an argument that at a wall, there is a boundary condition on a sound wave that the normal component of the gradient of the velocity potential must vanish

$$\hat{n} \cdot \vec{\nabla} \Phi = 0 \qquad ; \text{ at wall}$$

(b) Consider a one-dimensional sound wave propagating in a direction normal to a wall at $x = 0$. Show the following provides a solution to the wave equation for positive x

$$\Phi(x, t) = A \left[\cos k(x + ct) + \cos k(x - ct) \right] \qquad ; x \geq 0$$

(c) Show this solution obeys the appropriate boundary condition at the wall

$$\left[\frac{\partial \Phi(x,t)}{\partial x} \right]_{x=0} = -(kA) \left[\sin k(x + ct) + \sin k(x - ct) \right]_{x=0}$$
$$= 0$$

(d) Interpret the solution in part (b) in terms of an incident and reflected wave.

---

[4]Note that the cartesian basis vectors are constant under differentiation.

Solution to Problem 16.5

(a) The wall presents a constraint force on the fluid, which is held against the wall by its internal pressure. For an ideal fluid (and wall!), this constraint force will be normal to the wall. If the wall does not move, then the normal component of the fluid velocity must vanish there. Hence the boundary condition at the stationary wall is

$$\hat{n} \cdot \vec{v} = -\hat{n} \cdot \vec{\nabla}\Phi = 0 \qquad ; \text{at wall}$$

(b) Consider a one-dimensional sound wave propagating in the positive-x half-space, in a direction normal to the wall at $x = 0$. The following provides a solution to the wave equation for positive x, which will allow us to match the boundary condition at the wall [5]

$$\Phi(x,t) = A\left[\cos k(x + ct) + \cos k(x - ct)\right] \qquad ; x \geq 0$$

(c) It is readily established that this solution indeed obeys the appropriate boundary condition at the wall

$$-\left[\frac{\partial\Phi(x,t)}{\partial x}\right]_{x=0} = (kA)\left[\sin k(x + ct) + \sin k(x - ct)\right]_{x=0}$$
$$= 0$$

(d) The solution in part (b) consists of traveling waves of equal amplitude in the half-space moving both towards and away from the wall. This has the interpretation as an incident wave $A\cos k(x + ct)$ and a reflected wave $A\cos k(x - ct)$. At the wall with $x = 0$, the amplitude and phase of the reflected wave are unchanged from those of the incident wave.

Problem 16.6 Suppose the wall in Prob. 16.5 is oscillating with an angular frequency ω, so that the boundary condition at the wall now becomes

$$-\hat{n} \cdot \vec{\nabla}\Phi(x,t) = v_0 A \cos\omega t \qquad ; \text{at wall}$$

(a) Show the following driven, outgoing-wave solution to the wave equation satisfies this new boundary condition

$$\Phi(x,t) = -\frac{v_0}{k}A\sin\left[k(x - ct)\right] \qquad ; \omega = kc$$

---

[5]These are *plane waves*, with an identical disturbance in the entire transverse (y, z)-plane.

(b) The energy flux for irrotational isentropic fluid flow is given by Eq. (16.62) as

$$\vec{S}(\vec{x}, t) = -\rho \vec{\nabla} \Phi \frac{\partial \Phi(\vec{x}, t)}{\partial t}$$

Show the radiated energy flux from the oscillating wall is now given by

$$S_x^{\text{rad}}(x, t) = c(\rho v_0^2) \mathcal{A}^2 \cos^2 \left[k(x - ct) \right] \qquad ; \, \omega = kc$$

Interpret this result.

Solution to Problem 16.6

(a) A one-dimensional outgoing-wave solution to the wave equation for the velocity potential $\Phi(x, t)$ in the positive-x half-space is

$$\Phi(x, t) \propto \sin k(x - ct) \qquad\qquad ; \, \text{outgoing wave}$$

The boundary condition at the origin $(x = 0)$ with an oscillating wall is

$$-\hat{n} \cdot \vec{\nabla} \Phi(x, t) = v_0 \mathcal{A} \cos \omega t \qquad ; \, \text{at wall}$$

The solution that matches this boundary condition is evidently

$$\Phi(x, t) = -\frac{v_0}{k} \mathcal{A} \sin \left[k(x - ct) \right] \qquad ; \, kc = \omega$$

(b) The energy flux for irrotational isentropic fluid flow is given by Eq. (16.62) as

$$\vec{S}(\vec{x}, t) = -\rho \vec{\nabla} \Phi \frac{\partial \Phi(\vec{x}, t)}{\partial t}$$

It follows from this relation that the radiated energy flux in the x-direction from the oscillating wall is given by

$$S_x^{\text{rad}}(x, t) = c(\rho v_0^2) \mathcal{A}^2 \cos^2 \left[k(x - ct) \right] \qquad ; \, kc = \omega$$

As in Prob. 15.5, this is again the basic phenomenon of *radiation*. The oscillating wall puts energy into the system, here by doing work against the pressure in the fluid. The oscillating wall sends out a traveling sound wave that carries this energy. The energy can then be transported by the sound wave to another place, where the sound signal can then be detected.

Problem 16.7 (a) A piston under a pressure P expands quasistatically. Show that the reversible work done in the surroundings is $dW = PdV$ (see Fig. 16.1 below);

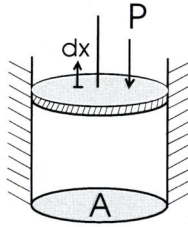

Fig. 16.1   A piston under a pressure P expands quasistatically.

(b) Hence, show that in the absence of any heat flow ($đQ = 0$), the first law of thermodynamics gives

$$dE = -dW = -PdV \qquad \text{; first law of thermo}$$
$$\text{; } đQ = 0$$

Solution to Problem 16.7

(a) Work is done by a force moving through a distance. The force the pressure exerts on the face of the piston is PA. If the piston is to move quasistatically through a distance dx, the gas inside must exert a pressure just infinitesimally greater than this. The work the system then does *on the external world* is

$$dW = PA\,dx = PdV \qquad \text{; on external world}$$

(b) The first law of thermodynamics states that the change in energy of the system E, a state function, is the difference of the heat flow in and the work flow out

$$dE = đQ - đW \qquad \text{; first law}$$

Hence, in the absence of any heat flow ($đQ = 0$), and with quasistatic reversible pressure-volume work, the first law of thermodynamics gives

$$dE = -dW = -PdV \qquad \text{; first law of thermodynamics}$$
$$\text{; } đQ = 0$$

****

(*Aside*) The second law of thermodynamics states that with quasistatic reversible heat flow, one has

$$dQ_R = T\,dS \qquad \text{; second law of thermodynamics}$$

where T is the absolute temperature and S is the *entropy*, another state function. Thus the first and second laws can be combined for quasistatic reversible processes to read

$$dE = T\, dS - P\, dV \qquad ; \text{ first and second laws of thermodynamics}$$

For *isentropic* processes, those with no reversible heat flow and constant entropy,

$$dE = -P\, dV \qquad ; \text{ isentropic processes}$$

****

Problem 16.8 The equation of state of a perfect gas is

$$PV = nRT \qquad ; \text{ perfect gas}$$

where n is the number of moles, T is the absolute temperature, and R is the gas constant[6]

$$R = 1.987 \text{ cal/mole-}°\text{K}$$

Furthermore, by the equipartition theorem, the internal energy is

$$E = \frac{3}{2}nRT$$

(a) Show the equation of state of the perfect gas, and its differential, can be written

$$PV = \frac{2}{3}E$$
$$P\, dV + V\, dP = \frac{2}{3}dE$$

(b) With no reversible heat flow (*isentropic*), the first law of thermodynamics says

$$dE = -dW = -PdV \qquad ; \text{ first law (isentropic)}$$

Hence, show

$$V\, dP = -\frac{5}{3}PdV$$

---

[6] Note $1\,\text{cal} = 4.184\,\text{joules}$.

Write the mass density of the gas as $\rho = M/V = nM_A/V$, where M_A is the molar atomic mass, and show

$$\frac{dP}{d\rho} = -\frac{5}{3}P\frac{dV}{Vd\rho} = \frac{5}{3}\frac{PV}{nM_A}$$

(c) Hence, show that the isentropic compressibility of the perfect gas is given by

$$\left(\frac{dP}{d\rho}\right)_S = \frac{5}{3}\frac{RT}{M_A} \qquad ; \text{ perfect gas}$$

Solution to Problem 16.8

It is assumed that you have seen the equation of state of a perfect gas

$$PV = nRT \qquad ; \text{ perfect gas}$$

The equipartition theorem of classical statistical mechanics states that there is a kinetic energy $k_BT/2$ per degree of freedom for each system where k_B is Boltzmann's constant, or $RT/2$ per mole.[7] The internal kinetic energy of a perfect gas is therefore

$$E = \frac{3}{2}nRT$$

(a) The above two expressions can be combined to express the equation of state of the perfect gas as

$$PV = \frac{2}{3}E \qquad ; \text{ perfect gas}$$

Differentiate this expression

$$P\,dV + V\,dP = \frac{2}{3}dE$$

(b) From the previous problem, with no reversible heat flow (*isentropic*), the first law of thermodynamics says

$$dE = -dW = -PdV$$

Hence,

$$V\,dP = -\frac{5}{3}PdV$$

---

[7]See, for example, [Walecka (2000)].

Write the mass density of the gas as $\rho = M/V = nM_A/V$, where M_A is the molar atomic mass.[8] It follows that

$$d\rho = -\frac{nM_A}{V^2}dV$$

Therefore

$$\frac{dP}{d\rho} = -\frac{5}{3}P\frac{dV}{Vd\rho} = \frac{5}{3}\frac{PV}{nM_A}$$

(c) Hence, the isentropic compressibility of the perfect gas is given by

$$\left(\frac{dP}{d\rho}\right)_S = \frac{5}{3}\frac{RT}{M_A} \qquad ; \text{ perfect gas}$$

Problem 16.9 Use the result in the previous problem to compute the speed of sound in air at $T = 273°\,$K from Eq. (16.53). Assume that air is 80% nitrogen (N_2) and 20% oxygen (O_2). Show

$$c_{\text{sound}} = 362 \text{ m/sec} \qquad ; \text{ calculated}$$

Compare with the experimental value in Prob. 16.2. Discuss.

Solution to Problem 16.9

Equation (16.53) gives the speed of irrotational, isentropic sound in a fluid as

$$\left(\frac{dP}{d\rho}\right)_S \equiv c^2$$

while Prob. 16.8 expresses the required compressibility in a perfect gas as

$$\left(\frac{dP}{d\rho}\right)_S = \frac{5}{3}\frac{RT}{M_A} \qquad ; \text{ perfect gas}$$

The required molar masses for air are readily obtained as

$$M_{O_2} = 32.00\,\text{gm}$$
$$M_{N_2} = 28.02\,\text{gm}$$

Then with air as 80% nitrogen (N_2) and 20% oxygen (O_2), the required "molar atomic mass" M_A in air is

$$M_A = [32.00 \times 0.2 + 28.02 \times 0.8]\,\text{gm} \qquad ; \text{ molar atomic mass}$$
$$= 28.816\,\text{gm}$$

---

[8]The mass of one mole of the constituents.

At $T = 273°$, the above then gives

$$c^2 = \frac{5 \times (1.987\,\text{cal/mole-°K}) \times (4.184\,\text{kg m}^2/\text{cal-s}^2) \times 273\,°\text{K}}{3 \times (28.816 \times 10^{-3}\,\text{kg/mole})}$$

$$= 1.312 \times 10^5\,\text{m}^2/\text{s}^2$$

Hence, the velocity of sound in a perfect gas is

$$c = 362\,\text{m/sec}$$

The experimental value of the speed of sound in air under these standard conditions is from Prob. 16.2

$$c_{\text{sound}} = 331\,\text{m/sec}$$

The above is pretty close. The difference is presumably due to the fact that air is not a perfect gas.

Problem 16.10 (a) In thermodynamics, one can make a Legendre transformation to a new thermodynamic variable, the *enthalpy*

$$H \equiv E + PV \qquad ; \text{enthalpy}$$

Use the first law of thermodynamics to show that for isentropic processes with no (reversible) heat flow

$$dH = VdP \qquad ; \text{isentropic process}$$

(b) Suppose one has a sample of mass M, volume V, and mass density $\rho = M/V$, and let $h = H/M$ be the enthalpy per unit mass. Show that for isentropic fluid flow, the force in Newton's second law is obtained from the negative gradient of the enthalpy[9]

$$-\vec{\nabla} h(\vec{x}, t) = -\frac{1}{\rho}\vec{\nabla} P(\vec{x}, t) \qquad ; \text{isentropic flow}$$

Solution to Problem 16.10

(a) Take the total differential of the enthalpy

$$dH = dE + PdV + VdP$$

Use the first law of thermodynamics for isentropic processes

$$dE = -PdV \qquad ; \text{first law, isentropic}$$

---

[9]Compare Eq. (16.45).

Hence

$$dH = V\,dP \qquad ; \text{ isentropic process}$$

(b) Suppose one has a sample of mass M, volume V, and mass density $\rho = M/V$, and let $h = H/M$ be the enthalpy per unit mass. Then

$$h = \epsilon + \frac{P}{\rho}$$

and from the above

$$dh = \frac{1}{\rho}dP$$

With a local dependence of the thermodynamic variables on $(\vec{x}, t)$, this relation gives

$$-\vec{\nabla}h(\vec{x}, t) = -\frac{1}{\rho}\vec{\nabla}P(\vec{x}, t) \qquad ; \text{ isentropic flow}$$

This is the force term in Newton's second law for fluid flow.

Problem 16.11 Give an explicit proof that with the hamiltonian (energy) density $\mathcal{H}(\vec{x}, t)$ of Eq. (16.12), and the energy flux $\vec{S}(\vec{x}, t)$ of Eqs. (16.32), one has the local statement of energy conservation

$$\frac{\partial \mathcal{H}(\vec{x}, t)}{\partial t} + \vec{\nabla} \cdot \vec{S}(\vec{x}, t) = 0$$

It is assumed here that $U = 0$ and there is no external force.[10]

Solution to Problem 16.11

We start with one of the required partial time derivatives

$$\frac{\partial}{\partial t}[\rho\epsilon(\rho)] = \frac{\partial \rho}{\partial t}\epsilon(\rho) + \rho\frac{d\epsilon(\rho)}{d\rho}\frac{\partial \rho}{\partial t}$$

Now make use of the continuity equation for the mass density

$$\frac{\partial \rho}{\partial t} + \vec{\nabla} \cdot (\rho\vec{v}) = 0 \qquad ; \text{ continuity eqn}$$

and the first law of thermodynamics for isentropic processes

$$\rho\frac{d\epsilon(\rho)}{d\rho} = \frac{P}{\rho} \qquad ; \text{ isentropic processes}$$

---

[10]This problem is more difficult (see [Fetter and Walecka (2003)]). *Viscosity*, where there is internal friction that converts kinetic flow energy into heat, is also absent here.

This gives

$$\frac{\partial}{\partial t}[\rho\epsilon(\rho)] = -\left[\vec{\nabla}\cdot(\rho\vec{v})\right]\left[\epsilon(\rho) + \frac{P}{\rho}\right]$$

We now write the r.h.s. as the divergence of the entire expression, and use Eq. (16.45) to eliminate the additional term[11]

$$\vec{\nabla}\left(\epsilon + \frac{P}{\rho}\right) = \frac{1}{\rho}\vec{\nabla}P \qquad ; \text{ isentropic flow}$$

This yields

$$\frac{\partial}{\partial t}[\rho\epsilon(\rho)] = -\vec{\nabla}\cdot\left[\rho\vec{v}\left(\epsilon + \frac{P}{\rho}\right)\right] + \vec{v}\cdot\vec{\nabla}P$$

Next, compute the partial time derivative of the second term

$$\frac{\partial}{\partial t}\left(\frac{1}{2}\rho v^2\right) = \frac{\partial}{\partial t}\left(\frac{1}{2}\rho\,\vec{v}\cdot\vec{v}\right)$$

$$= \frac{\partial\rho}{\partial t}\left(\frac{1}{2}v^2\right) + \rho\vec{v}\cdot\frac{\partial\vec{v}}{\partial t}$$

Use Newton's second law for irrotational flow in Eq. (16.43)[12]

$$\frac{\partial\vec{v}}{\partial t} + \vec{\nabla}\left(\frac{1}{2}v^2\right) = -\frac{1}{\rho}\vec{\nabla}P \qquad ; \text{ irrotational flow}$$

and the continuity equation again, to obtain

$$\frac{\partial}{\partial t}\left(\frac{1}{2}\rho v^2\right) = -\left[\vec{\nabla}\cdot(\rho\vec{v})\right]\left(\frac{1}{2}v^2\right) - \rho\vec{v}\cdot\left[\vec{\nabla}\left(\frac{1}{2}v^2\right) + \frac{1}{\rho}\vec{\nabla}P\right]$$

$$= -\vec{\nabla}\cdot\left[\rho\vec{v}\left(\frac{1}{2}v^2\right)\right] - \vec{v}\cdot\vec{\nabla}P$$

Finally, add the two time derivatives together to arrive at the desired result

$$\frac{\partial\mathcal{H}(\vec{x},t)}{\partial t} + \vec{\nabla}\cdot\vec{S}(\vec{x},t) = 0$$

$$\mathcal{H}(\vec{x},t) = \frac{1}{2}\rho v^2 + \rho\epsilon(\rho)$$

$$\vec{S}(\vec{x},t) = \rho\vec{v}\left[\frac{1}{2}v^2 + \epsilon(\rho) + \frac{P}{\rho}\right]$$

---

[11]Note the general vector relation $\vec{\nabla}\cdot(a\vec{b}) = a(\vec{\nabla}\cdot\vec{b}) + \vec{b}\cdot(\vec{\nabla}a)$.

[12]Remember $U = 0$.

This is the conservation law for the energy density for irrotational, isentropic fluid flow in the absence of an external force ($U = 0$). Here $\mathcal{H}(\vec{x}, t)$ is the hamiltonian (energy) density, and $\vec{S}(\vec{x}, t)$ is the energy flux. A direct calculation of this relation from the conservation of the energy-momentum tensor in the lagrangian mechanics of fluid flow is obtained in the next problem.

Problem 16.12 (a) Introduce the same notation as in Sec. 15.6, only now with three spatial coordinates so that $x_\mu = (x, y, z, ict)$. Show that Lagrange's equations for irrotational, isentropic flow become

$$\frac{\partial}{\partial x_\mu} \left[\frac{\partial \mathcal{L}}{\partial(\partial \rho/\partial x_\mu)} \right] - \frac{\partial \mathcal{L}}{\partial \rho} = 0$$

$$\frac{\partial}{\partial x_\mu} \left[\frac{\partial \mathcal{L}}{\partial(\partial \Phi/\partial x_\mu)} \right] - \frac{\partial \mathcal{L}}{\partial \Phi} = 0$$

(b) Since there is no dependence on $\partial \rho/\partial x_\mu$ in the lagrangian density in Eq. (16.16), show the energy-momentum tensor takes the form

$$T_{\mu\nu} \equiv \mathcal{L}\delta_{\mu\nu} - \frac{\partial \mathcal{L}}{\partial(\partial \Phi/\partial x_\mu)} \frac{\partial \Phi}{\partial x_\nu}$$

$$; \text{ irrotational isentropic flow}$$

(c) Assume no external force, and write out the conservation law in Eq. (15.68) for $\nu = 4$

$$\frac{\partial T_{\mu 4}}{\partial x_\mu} = 0$$

Show this reproduces the result in Prob. 16.11;

(d) Focus on the spatial indices (i, j). Show that *along the actual path* of the motion

$$T_{ij} = P\delta_{ij} + \rho v_i v_j \qquad ; \text{ along path}$$

Solution to Problem 16.12

(a) There are two generalized coordinates (ρ, Φ). It then follows from Eqs. (10.25), (16.19), and Prob. 16.4 that one has a Lagrange equation of the following form for each generalized coordinate

$$\frac{\partial}{\partial x} \left[\frac{\partial \mathcal{L}}{\partial(\partial \rho/\partial x)} \right] + \frac{\partial}{\partial y} \left[\frac{\partial \mathcal{L}}{\partial(\partial \rho/\partial y)} \right] + \frac{\partial}{\partial z} \left[\frac{\partial \mathcal{L}}{\partial(\partial \rho/\partial z)} \right] + \frac{\partial}{\partial t} \left[\frac{\partial \mathcal{L}}{\partial(\partial \rho/\partial t)} \right]$$

$$= \frac{\partial \mathcal{L}}{\partial \rho}$$

In the new notation, these equations read

$$\frac{\partial}{\partial x_\mu}\left[\frac{\partial\mathcal{L}}{\partial(\partial\rho/\partial x_\mu)}\right] - \frac{\partial\mathcal{L}}{\partial\rho} = 0$$

$$\frac{\partial}{\partial x_\mu}\left[\frac{\partial\mathcal{L}}{\partial(\partial\Phi/\partial x_\mu)}\right] - \frac{\partial\mathcal{L}}{\partial\Phi} = 0$$

(b) Equation (15.72) states that the corresponding energy-momentum tensor is

$$\mathcal{T}_{\mu\nu} \equiv \mathcal{L}\delta_{\mu\nu} - \frac{\partial\mathcal{L}}{\partial(\partial\Phi/\partial x_\mu)}\frac{\partial\Phi}{\partial x_\nu} - \frac{\partial\mathcal{L}}{\partial(\partial\rho/\partial x_\mu)}\frac{\partial\rho}{\partial x_\nu}$$

The lagrangian density for irrotational isentropic fluid flow in Eq. (16.16) is

$$\mathcal{L}\left(\frac{\partial\Phi}{\partial t},\vec{\nabla}\Phi,\rho;\vec{x}\right) = \rho\frac{\partial\Phi}{\partial t} - \frac{1}{2}\rho(\vec{\nabla}\Phi)^2 - \rho\epsilon(\rho) - \rho U(\vec{x})$$

Since there is no dependence on $\partial\rho/\partial x_\mu$, the energy-momentum tensor takes the form

$$\mathcal{T}_{\mu\nu} \equiv \mathcal{L}\delta_{\mu\nu} - \frac{\partial\mathcal{L}}{\partial(\partial\Phi/\partial x_\mu)}\frac{\partial\Phi}{\partial x_\nu}$$

(1) Consider $\mathcal{T}_{44}(\vec{x},t)$

$$\mathcal{T}_{44}(\vec{x},t) = \mathcal{L} - \left[\frac{\partial\mathcal{L}}{\partial(\partial\Phi/\partial t)}\right]\frac{\partial\Phi(\vec{x},t)}{\partial t}$$

$$= \mathcal{L} - \Pi_\Phi(\vec{x},t)\frac{\partial\Phi(\vec{x},t)}{\partial t}$$

$$= -\mathcal{H}(\vec{x},t)$$

This is the negative of the hamiltonian (energy) density in Eq. (16.12)

$$\mathcal{H}(\vec{x},t) = \mathcal{T} + \mathcal{V} \qquad\qquad ; \text{ hamiltonian density}$$

$$= \frac{1}{2}\rho(\vec{\nabla}\Phi)^2 + \rho\epsilon(\rho) + \rho U(\vec{x})$$

(2) Consider $\mathcal{T}_{j4}(\vec{x},t)$ where j is a spatial index

$$\mathcal{T}_{j4}(\vec{x},t) = \frac{i}{c}\left[\frac{\partial\mathcal{L}}{\partial(\partial\Phi/\partial x_j)}\right]\frac{\partial\Phi(\vec{x},t)}{\partial t}$$

$$= \frac{i}{c}S_j(\vec{x},t)$$

where $\vec{S}(\vec{x}, t)$ is the energy flux of Eqs. (16.31)–(16.32)

$$\vec{S}(\vec{x}, t) = -\rho \vec{\nabla} \Phi \frac{\partial \Phi(\vec{x}, t)}{\partial t} \qquad ; \text{ energy flux}$$

$$= \rho \vec{v} \left[\frac{1}{2} (\vec{\nabla} \Phi)^2 + \epsilon(\rho) + \frac{P}{\rho} + U(\vec{x}) \right]$$

(c) Assume no external force and $U(\vec{x}) = 0$. If the lagrangian density $\mathcal{L}$ has no *explicit* dependence on x_μ, then the conservation law in Eq. (15.68) says that

$$\frac{\partial T_{\mu\nu}}{\partial x_\mu} = 0$$

Apply this for $\nu = 4$, and use the above,

$$\frac{\partial T_{\mu 4}}{\partial x_\mu} = \frac{i}{c} \left[\frac{\partial \mathcal{H}(\vec{x}, t)}{\partial t} + \vec{\nabla} \cdot \vec{S}(\vec{x}, t) \right] = 0$$

This reproduces the result in Prob. 16.11.

(d) For two spatial indices (i, j)

$$T_{ij} = \mathcal{L} \delta_{ij} - \frac{\partial \mathcal{L}}{\partial(\partial \Phi / \partial x_i)} \frac{\partial \Phi}{\partial x_j}$$

$$= \mathcal{L} \delta_{ij} + \rho v_i v_j$$

Along the actual path of the motion, we can use Bernoulli's theorem in Eq. (16.18),[13]

$$\frac{\partial \Phi}{\partial t} = \frac{1}{2} (\vec{\nabla} \Phi)^2 + \epsilon(\rho) + \frac{P}{\rho} + U(\vec{x})$$

Hence, along the actual path of the motion,

$$T_{ij} = P \delta_{ij} + \rho v_i v_j \qquad ; \text{ along path}$$

****

(*Aside*) Let us see what $\nu = j$, a spatial index, gives us for the conservation of the energy-momentum tensor. One has

$$T_{4j} = -\frac{\partial \mathcal{L}}{\partial(\partial \Phi / \partial x_4)} \frac{\partial \Phi}{\partial x_j} = ic \frac{\partial \mathcal{L}}{\partial(\partial \Phi / \partial t)} v_j = ic P_j$$

---

[13] *Warning:* You cannot substitute the actual path back into the lagrangian before you start, since Hamilton's principle forces you to consider *variations* about the actual path.

where $\mathcal{P}_j$ is the jth component of the explicit *momentum density*

$$\mathcal{P}_j \equiv \rho v_j \qquad\qquad ; \text{ momentum density}$$

The conservation law for the energy-momentum tensor then states that

$$\frac{\partial T_{\mu j}}{\partial x_\mu} = \frac{\partial}{\partial t}(\rho v_j) + \frac{\partial}{\partial x_i}(P\delta_{ij} + v_i v_j \rho) = 0$$

which can be rewritten as

$$\frac{\partial \mathcal{P}_j}{\partial t} + \vec{\nabla} \cdot (\vec{v}\,\mathcal{P}_j) = -\frac{\partial P}{\partial x_j}$$

Take the time derivative, and make use of the continuity equation for the mass density, to arrive at

$$\frac{\partial v_j}{\partial t} + \left(\vec{v} \cdot \vec{\nabla}\right) v_j = -\frac{1}{\rho}\frac{\partial P}{\partial x_j}$$

In vector form, this is just Newton's second law for fluid flow in Eq. (16.40), with no external force

$$\frac{\partial \vec{v}}{\partial t} + \left(\vec{v} \cdot \vec{\nabla}\right)\vec{v} = -\frac{1}{\rho}\vec{\nabla}P \qquad\qquad ; \text{ Newton's law}$$

****

Problem 16.13 A moving wall sends a plane-wave sound *pulse* through a medium of the form examined in Prob. 14.1,[14]

$$\Phi(x,t) = Ae^{-(x-ct)^2/a^2} \qquad ; \text{ sound pulse}$$

(a) Show from Eq. (16.55) that the mass density variation is

$$\delta\rho(x,t) = \rho_0 \left(\frac{2A}{ca}\right)\left[\frac{(x-ct)}{a}\right]e^{-(x-ct)^2/a^2}$$

(b) Show from Eq. (16.66) that the corresponding fluid velocity is

$$v_x(x,t) = c\left(\frac{2A}{ca}\right)\left[\frac{(x-ct)}{a}\right]e^{-(x-ct)^2/a^2}$$

(c) Show from Eq. (16.68) that the energy flux is

$$S_x(x,t) = \rho_0 c^3 \left(\frac{2A}{ca}\right)^2\left[\frac{(x-ct)}{a}\right]^2 e^{-2(x-ct)^2/a^2}$$

(d) Sketch and interpret these results.

---

[14]Recall Fig. 14.1 in the text.

Solution to Problem 16.13

The solution to Prob. 15.2 exhibits the displacement and momentum density for a traveling wave on a string. This problem displays similar quantities for a traveling sound impulse in the fluid. The velocity potential is given as

$$\Phi(x,t) = Ae^{-(x-ct)^2/a^2} \qquad ; \text{ sound pulse}$$

The fluid velocity and density variation are calculated immediately from Eqs. (16.61) and (16.66)

$$\frac{1}{\rho_0(2A/ca)}\delta\rho(x,t) = \left[\frac{(x-ct)}{a}\right]e^{-(x-ct)^2/a^2}$$

$$\frac{1}{c(2A/ca)}v_x(x,t) = \left[\frac{(x-ct)}{a}\right]e^{-(x-ct)^2/a^2}$$

The energy flux follows from Eq. (16.68)

$$\frac{1}{\rho_0 c^3(2A/ca)^2}S_x(x,t) = \left[\frac{(x-ct)}{a}\right]^2 e^{-2(x-ct)^2/a^2}$$

We sketch these quantities in Fig. 16.2 below.

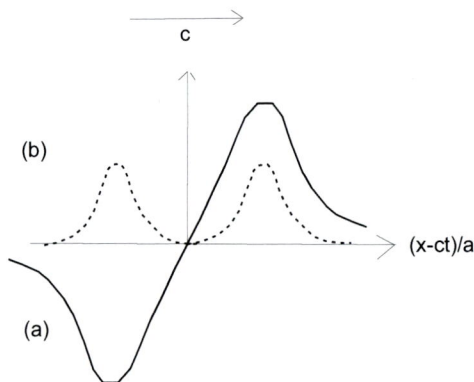

Fig. 16.2   Sketch of sound pulse $\Phi(x,t) = Ae^{-(x-ct)^2/a^2}$ in a fluid: (a) Density variation and fluid velocity $\delta\rho(x,t)/[\rho_0(2A/ca)]$ and $v_x(x,t)/[c(2A/ca)]$; and (b) Energy flux $S_x(x,t)/[\rho_0 c^3(2A/ca)^2]$.

It is interesting how similar the energy flux here is to the energy flux on the string in Fig. 15.1 shown in the solution to Prob. 15.2.

Appendix A

Numerical Methods

Problem A.1 Show that the lagrangian for the planer motion of a particle with mass m in the point gravitational potential $V(r) = -mMG/r$, with generalized coordinates (r, θ), is

$$L(\dot{r}, \dot{\theta}, r, \theta) = \frac{m}{2}\left(\dot{r}^2 + r^2\dot{\theta}^2\right) + \frac{mMG}{r}$$

Use this to derive Lagrange's Eqs. (A.1) in the text.

Solution to Problem A.1

The kinetic energy in polar coordinates is given in Eq. (9.26), and the potential energy with a point source in Eq. (6.12). Hence the lagrangian is

$$L(\dot{r}, \dot{\theta}, r, \theta) = T - V = \frac{m}{2}\left(\dot{r}^2 + r^2\dot{\theta}^2\right) + \frac{mMG}{r}$$

Lagrange's Eqs. (9.31) then give

$$\frac{d^2r}{dt^2} = \frac{l^2}{m^2r^3} - \frac{MG}{r^2} \qquad ; \; l^2 = (mr^2\dot{\theta})^2$$

$$\frac{d}{dt}\left(mr^2\dot{\theta}\right) = 0$$

Here l^2 is the square of the angular momentum.

Problem A.2 Verify Eq. (A.5), and then justify it.

Solution to Problem A.2

We start with the justification. We want to convert the second-order differential equation in Eqs. (A.1) to a finite-difference matrix equation, since it is a simple task for computers to iterate such equations. Introduce the column vector

$$\underline{Z}_n \equiv \begin{bmatrix} r_n \\ \dot{r}_n \\ \theta_n \end{bmatrix}$$

We do not also need $\dot{\theta}$ here, since we have the conserved angular momentum $l = mr^2\dot{\theta}$, which must be specified. The components of $\underline{Z}_n$ will be denoted $(Z_n)_i$ with $i = (1, 2, 3)$.

If Δt indicates a small step in time, which can be decreased while increasing the number of iteration steps, then the finite-difference equation reads

$$\underline{Z}_{n+1} = \underline{Z}_n + \Delta t \begin{bmatrix} \dot{r}_n \\ \ddot{r}_n \\ \dot{\theta}_n \end{bmatrix}$$

What makes this work is that the quantities in $\ddot{r}_n$ can be related back to those in $\underline{Z}_n$ through the differential equation. Thus

$$\begin{bmatrix} \dot{r}_n \\ \ddot{r}_n \\ \theta_n \end{bmatrix} = \begin{bmatrix} (Z_n)_2 \\ l^2/m^2(Z_n)_1^3 - MG/(Z_n)_1^2 \\ l/m(Z_n)_1^2 \end{bmatrix}$$

It follows that

$$\underline{Z}_{n+1} = \underline{Z}_n + \Delta t \begin{bmatrix} (Z_n)_2 \\ l^2/m^2(Z_n)_1^3 - MG/(Z_n)_1^2 \\ l/m(Z_n)_1^2 \end{bmatrix} \qquad ; \text{ given } \underline{Z}_1$$

The initial condition $\underline{Z}_1$ must also be specified.

Now just apply this analysis to the dimensionless form of the orbit equations in Appendix A. Introduce

$$\underline{Z}_n \equiv \begin{bmatrix} u_n \\ (du/d\sigma)_n \\ \theta_n \end{bmatrix}$$

Newton's laws can then be recast as the following dimensionless, finite-difference, matrix equation

$$\underline{Z}_{n+1} = \underline{Z}_n + \Delta \begin{bmatrix} (Z_n)_2 \\ L^2/(Z_n)_1^3 - 1/2(Z_n)_1^2 \\ -L/(Z_n)_1^2 \end{bmatrix} \qquad ; \text{ given } \underline{Z}_1$$

The minus sign in the last entry occurs because the calculation in Fig. A.1 in the text uses the spherical polar angle measured with respect to the z-axis (in plane) instead of the plane polar angle measured with respect to the x-axis; the spherical polar angle *decreases* as one moves up along the trajectory.

Problem A.3 (a) Use the program mentioned in Appendix A, or write your own, and generate some additional scattering orbits for a particle in the given point gravitational potential;

(b) Generate some bound (closed) orbits in that potential.

Solution to Problem A.3

(a) We did use the Mathcad program mentioned in Appendix A to generate some additional scattering orbits, and also some *closed* orbits.

(b) We include here a figure of one of the closed orbits. The initial conditions for the representative trajectory in Fig. A.1 below are

$$\underline{Z}_1 = \begin{bmatrix} 12 \\ -0.05 \\ 2 \end{bmatrix} \qquad ; L = 1.20 \qquad (A.1)$$

Here $L = 1.20$ was adjusted to close the orbit. The particle again moves up along the trajectory.

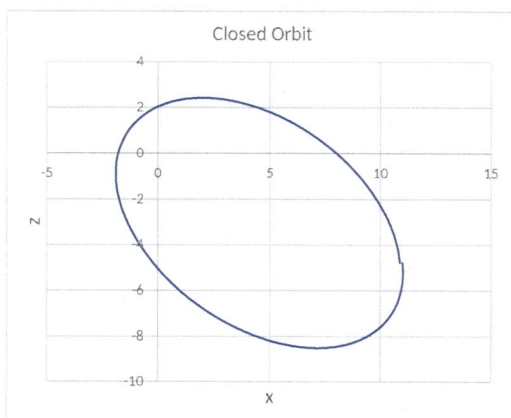

Fig. A.1 A closed orbit obtained through numerical iteration of the dimensionless finite difference equations using Mathcad 7, with the above initial conditions. We employed $\Delta = 0.0057$, with $N = 30,000$ iterations. The figure was prepared using Excel.

Problem A.4 Since the speed of light c does not appear in the equations of motion in Eqs. (A.1), show that *any* velocity $\bar{c}$ can be used in going to the dimensionless Newtonian orbit results in Appendix A and Prob. A.3. Discuss the implications of this observation.

Solution to Problem A.4

One ordinarily characterizes the scattering orbit illustrated in Fig. A.1 in the text by the physical *impact parameter* and *incident velocity* $(x_p, \dot{z}_p)$.[1] The particular orbit shown in that figure corresponds to a given value of this pair. By choosing another velocity $\bar{c}$ to be used in going to dimensionless variables, the *curve* does not change, but one is describing another set of values of this physical pair. So, in fact, the single orbit in Fig. A.1 in the text describes *all* of the orbits along a path in the $(x_p, \dot{z}_p)$-plane.

It is clear from the definition of the dimensionless variables in Eqs. (A.4) that as we go to the new velocity $\bar{c}$:

- Physical distances scale up from dimensionless distances by a factor of $(c/\bar{c})^2$;
- Physical velocities scale down from dimensionless velocities by $(\bar{c}/c)$.

In choosing a $\bar{c}$ to work with, we must ensure that:

- The distance of closest approach to the source is greater than its actual radius;
- The incident particle velocity is less than the speed of light.

---

[1] Or impact parameter and incident *momentum*.

Appendix B

Significant Names in Classical Mechanics

Galileo Galilei (1564–1642)
Johannes Kepler (1571–1630)
Isaac Newton (1642–1726)
Brook Taylor (1685–1731)
Daniel Bernoulli (1700–1782)
Leonhard Euler (1707–1783)
Jean Le Rond D'Alembert (1717–1783)
Charles Augustin de Coulomb (1736–1806)
Joseph Louis Lagrange (1736–1813)
Pierre Simon de Laplace (1749–1827)
Adrien Marie Legendre (1752–1833)
William Rowan Hamilton (1805–1865)
Joseph Liouville (1809–1882)
James Prescott Joule (1818–1889)
William Thomson (Lord Kelvin) (1824–1907)
Hendrik Lorentz (1853–1928)
Max Karl Ludwig Planck (1858–1947)
Karl Schwarzschild (1873–1916)
Albert Einstein (1879–1955)
Niels Henrik David Bohr (1885–1962)

Bibliography

Amore, P., and Walecka, J. D., (2013). *Introduction to Modern Physics: Solutions to Problems*, World Scientific Publishing Company, Singapore

Amore, P., and Walecka, J. D., (2014). *Topics in Modern Physics: Solutions to Problems*, World Scientific Publishing Company, Singapore

Amore, P., and Walecka, J. D., (2015). *Advanced Modern Physics: Solutions to Problems*, World Scientific Publishing Company, Singapore

Fetter, A. L. and Walecka, J. D., (2003). *Theoretical Mechanics of Particles and Continua*, McGraw-Hill, New York (1980); reissued by Dover Publications, Mineola, New York

Fetter, A. L. and Walecka, J. D., (2003a). *Quantum Theory of Many-Particle Systems*, McGraw-Hill, New York (1971); reissued by Dover Publications, Mineola, New York

Fetter, A. L., and Walecka, J. D., (2006). *Nonlinear Mechanics: A Supplement to Theoretical Mechanics of Particles and Continua*, Dover Publications, Mineola, New York

Freedman, R., Ruskell, T., Keston, P. M., and Tauck, D. L., (2013). *College Physics*, W. H. Freeman, San Francisco, CA

Goldstein, H., Poole, C. P., and Safko, J., (2011). *Classical Mechanics*, 3rd international economy ed., Pearson Education, London, UK

Halliday, D., Resnick, R., and Walker, J., (2013). *Fundamentals of Physics, 10th ed.*, J. Wiley and Sons, New York, NY

Kibble, T. W. B., and Berkshire, F. H., (2004). *Classical Mechanics, 5th ed.*, Imperial College, London, UK

Kleppner, D., and Kolenkow, R., (2013). *An Introduction to Mechanics, 2nd ed.*, Cambridge University Press, Cambridge, UK

Landau, L. D., and Lifshitz, E. M., (1976). *Mechanics*, Butterworth-Heinemann, Oxford, UK

Morin, D., (2008). *Introduction to Classical Mechanics: With Problems and Solutions*, Cambridge University Press, Cambridge, UK

Ohanian, H. C., (1985). *Physics*, W.W. Norton and Co., New York, NY

Ohanian, H. C., (1995). *Modern Physics, 2nd ed.*, Prentice-Hall, Upper Saddle River, NJ

Taylor, J. R., (2004). *Classical Mechanics*, University Science Books, Sausilito, CA

Thornton, S. T., and Marion, J. B., (2012). *Classical Dynamics of Particles and Systems, 5th ed.*, Cengage Learning, Boston, MA

Walecka, J. D., (2000). *Fundamentals of Statistical Mechanics: Manuscript and Notes of Felix Bloch, prepared by J. D. Walecka*, World Scientific Publishing Company, Singapore; originally published by Stanford University Press, Stanford, CA (1989)

Walecka, J. D., (2004). *Theoretical Nuclear and Subnuclear Physics, 2nd ed.*, World Scientific, Singapore

Walecka, J. D., (2007). *Introduction to General Relativity*, World Scientific Publishing Company, Singapore

Walecka, J. D., (2008). *Introduction to Modern Physics: Theoretical Foundations*, World Scientific Publishing Company, Singapore

Walecka, J. D., (2010). *Advanced Modern Physics: Theoretical Foundations*, World Scientific Publishing Company, Singapore

Walecka, J. D., (2011). *Introduction to Statistical Mechanics*, World Scientific Publishing Company, Singapore

Walecka, J. D., (2013). *Topics in Modern Physics: Theoretical Foundations*, World Scientific Publishing Company, Singapore

Walecka, J. D., (2017). *Introduction to Statistical Mechanics: Solutions to Problems*, World Scientific Publishing Company, Singapore

Walecka, J. D., (2017a). *Introduction to General Relativity: Solutions to Problems*, World Scientific Publishing Company, Singapore

Walecka, J. D., (2018). *Introduction to Electricity and Magnetism*, World Scientific Publishing Company, Singapore

Walecka, J. D., (2019). *Introduction to Electricity and Magnetism: Solutions to Problems*, World Scientific Publishing Company, Singapore

Walecka, J. D., (2020). *Introduction to Classical Mechanics*, World Scientific Publishing Company, Singapore

Wiki (2019). *The Wikipedia*, http://en.wikipedia.org/wiki/(topic)

Index